Operations Research

The proposed book volume covers cutting-edge research that includes operations research and decision sciences. It examines finite queueing systems with unreliable service provisioning and feedback strategies. Retrial queueing systems with customer feedback and individual abandonments are dynamically investigated. The book presents fuzzy logic approaches for multi-objective transportation problems. Multi-objective programming methodologies effectively deal with problems associated with catastrophe management. Healthcare decision-making under uncertain contexts deserves special attention. Inventory models combine trade credit financing with credit-linked demand functions. Financial risk analysis, business estimation, and bilevel portfolio optimization strategies are thoroughly investigated.

Features:

- Delves into practical approaches to operational research concerns, offering frameworks that address real-world scenarios.
- Introduces new techniques and strategies for problem-solving, involving the development of physical scenarios, effective algorithms, and straightforward implementation.
- Discusses topics such as inventory control models, root mean square optimization, fuzzy optimization with unique membership functions, and traffic simulation for telecommunication networks.
- Covers mathematical perspectives, such as variable costs, perishable and nonperishable products for consumers, fuzzy decision variables, and constraint qualifications.
- Presents an in-depth analysis of the latest developments shaping the field of operations research.

It is primarily written for graduate students and academic researchers in operations research, industrial engineering, systems engineering, production engineering, manufacturing engineering, mathematics, and quality control.

Advanced Research in Reliability and System Assurance Engineering

Series Editor: Mangey Ram, Professor, Graphic Era University, Uttarakhand, India

Operations Research
Methods, Techniques, and Advancements
Edited by Amit Kumar and Mangey Ram

Statistical Modeling of Reliability Structures and Industrial Processes
Edited by Ioannis S. Triantafyllou and Mangey Ram

Industrial Reliability and Safety Engineering
Applications and Practices
Edited by Dilbagh Panchal, Mangey Ram, Prasenjit Chatterjee, and Anish Kumar Sachdeva

Reliability and Maintenance Modeling with Optimization
Advances and Applications
Edited by Mitsutaka Kimura, Satoshi Mizutani, Mitsuhiro Imaizumi, and Kodo Ito

Intelligent Prognostics for Engineering Systems with Machine Learning Techniques
Edited by Gunjan Soni, Om Prakash Yadav, Gaurav Kumar Badhotiya, and Mangey Ram

Statistical Methods and Applications in Systems Assurance and Quality
Edited by Sotiris Bersimis, Polychronis Economou, and Athanasios Rakitzis

Designing Faultless Mechanical Products Based on Advanced Reliability Analysis
Seongwoo Woo

Operations Research
Recent Advances and Applications
Edited by Shreekant Varshney, Jyotiranjan Nayak, Mayank Gupta, Chandra Shekhar, and Balakrushna Tripathy

For more information about this series, please visit: https://www.routledge.com/Advanced-Research-in-Reliability-and-System-Assurance-Engineering/book-series/CRCARRSAE

Operations Research
Recent Advances and Applications

Edited by Shreekant Varshney,
Jyotiranjan Nayak, Mayank Gupta,
Chandra Shekhar, and Balakrushna Tripathy

CRC Press
Taylor & Francis Group
Boca Raton London New York

CRC Press is an imprint of the
Taylor & Francis Group, an **informa** business

Designed cover image: © Chandra Shekhar

First edition published 2026
by CRC Press
2385 NW Executive Center Drive, Suite 320, Boca Raton FL 33431

and by CRC Press
4 Park Square, Milton Park, Abingdon, Oxon, OX14 4RN

CRC Press is an imprint of Taylor & Francis Group, LLC

ISBN: 978-1-041-07761-9 (hbk)
ISBN: 978-1-041-07765-7 (pbk)
ISBN: 978-1-003-64209-1 (ebk)

DOI: 10.1201/9781003642091

Typeset in Sabon
by Apex CoVantage, LLC

Contents

Preface

This proposed volume delves into practical approaches to operational research concerns, offering frameworks that address real-world scenarios. It introduces several new techniques and strategies for problem-solving, involving the development of physical scenarios, effective algorithms, and straightforward implementation. The authenticity of outcomes is ensured through specific objectives, accurate input data, comprehensive verification, and efficient procedures. Among the diverse topics covered in this collection of chapters are inventory control models, root mean square (RMS) optimization, fuzzy optimization with unique membership functions, traffic simulation for telecommunication networks, stochastic optimization, and many more. Addressing real-world problems characterized by uncertainty, unpredictability, and complex dynamics necessitates the application of stochastic optimization. Unlike deterministic optimization, stochastic optimization considers probabilistic factors, making it suitable for situations where parameters are subject to random fluctuations or when accurate information is not available.

Stochastic optimization aids decision-makers across various sectors, including finance, logistics, and telecommunications, in modeling dynamic environments, accounting for uncertainty, and optimally allocating resources where ambiguity exists. Organizations can make robust, flexible decisions in complex real-world scenarios due to their ability to manage unpredictability and uncertainty, ultimately enhancing productivity, resilience, and competitiveness. The impact on the global economy has been studied and elucidated, establishing a framework for future researchers to investigate and predict factors influencing financial system growth and stability. Various mathematical perspectives, such as variable costs, perishable and nonperishable products for consumers, fuzzy decision variables, and numerous constraint qualifications, are employed to elucidate inventory control problems.

Traffic simulation is crucial in telecommunication networks for forecasting and controlling the dynamics of data flow. By simulating real-world scenarios, operators can enhance resource allocation, identify congestion points, and maximize network performance. This facilitates capacity planning, service provisioning, and infrastructure design through complex algorithms and modeling methodologies, ensuring reliable and effective communication services. In conclusion, it enables telecommunication firms to provide lag-free connectivity and address the ever-evolving requirements of modern digital communication.

The robust statistical method known as response surface methodology (RSM) finds applications in various domains such as engineering, energy systems, and statistical optimization. Its effectiveness lies in its ability to investigate the complex interrelationships between several input parameters and their impact on an output response. RSM enhances the understanding of underlying processes and their optimization for desired outcomes by systematically altering input components and tracking corresponding changes in the response. Engineers and researchers employ this method to identify the most appropriate strategies for achieving the desired performance or maximum output by developing predictive models that reflect the nonlinear interactions between decision variables. RSM provides a structured method for testing and optimization, resulting in reduced expenses, improved product quality, and increased efficiency. It can be used to optimize manufacturing processes as well as product formulations and architectures. Due to its effectiveness and flexibility, RSM is an essential tool for solving problems and making decisions across a variety of sectors. In other models, the most effective approach to selecting suitable parameters is determined through the implementation of metaheuristics. The queueing model evaluates a service system and suggests appropriate algorithm formulations.

Shreekant Varshney
Gandhinagar, India
Jyotiranjan Nayak
Hyderabad, India
Mayank Gupta
Gandhinagar, India
Chandra Shekhar
Pilani, India
Balakrushna Tripathy
Vellore, India
December 2025

Acknowledgments

नमस्ते सदा वत्सले मातृभूमे
त्वया हिन्दुभूमे सुखं वर्धितोऽहम्।
महामङ्गले पुण्यभूमे त्वदर्थे
पतत्वेष कायो नमस्ते नमस्ते।।

This book is transcribed while the editors are supported by institutes of repute, Pandit Deendayal Energy University (PDEU), IFHE, Birla Institute of Technology and Science (BITS) Pilani, and VIT Vellore. We acknowledge the heartiest thanks to these institutes' administrative and academic support. Very special thanks are offered to Prof. Mangey Ram, series editor, and Mr. Gauravjeet Singh, senior commissioning editor, CRC Press, for their support and suggestions to improve the text. We are also indebted to the learned reviewers, whose remarks were constructive and who thoroughly improved the chapters' manuscripts. The contributors of a variety of manuscripts related to operations research and allied fields deserve a note of thanks for sharing content-rich thoughts. We are grateful to all well-wishers who enlightened the path of progress.

We thank our parents, children, friends, and society for inspiration, encouragement, and cooperation. Dr. Varshney dedicates this to his grandparents, Shri C. B. Varshney and Smt. Rammurti Devi. Prof. Nayak is grateful to his wife, Ms. Namita, for her prompt support. Dr. Gupta expresses gratitude for the support of his wife and parents. Prof. Shekhar acknowledges the support of his father, Shri L. P. Jayaswal, and wife, Ms. Shikha Gupta, for their understanding during busy times and continued motivation for the book's preparation, and Prof. Tripathy expresses gratitude to all the supporters for their prompt support.

Lastly, we are also thankful to all well-wishers who provided us with the best possible support, help, and backup in completing the work.

Shreekant Varshney
Jyotiranjan Nayak
Mayank Gupta
Chandra Shekhar
Balakrushna Tripathy

About the Editors

Shreekant Varshney is an assistant professor in the Department of Mathematics, School of Technology (SoT), Pandit Deendayal Energy University (PDEU), Gandhinagar, Gujarat, India. Prior to joining PDEU, he worked as an assistant professor in the Department of Mathematics at the Faculty of Science and Technology (ICFAI Tech), IFHE, Hyderabad, India. He obtained his PhD from the Birla Institute of Technology and Science (BITS), Pilani, Pilani Campus, in 2020. Also, he has cleared CSIR JRF/NET twice with AIR 18 and 67, respectively. He is actively involved in the research domains, namely, queueing theory, machine repair problem (MRP), optimal control, reliability and maintainability, stochastic modeling, electric vehicle (EV) network modeling, cost optimization, sensitivity analysis, evolutionary computation, statistical analysis, and fuzzy set and logic. He has taught undergraduate and postgraduate subjects, namely, linear algebra, differential equations, complex analysis, advanced calculus, numerical methods, stochastic processes, differential geometry, and probability and statistics. He also has experience teaching courses like research methodology, stochastic simulation, and introduction to queueing theory for PhD students. Besides attending, presenting scientific papers, and delivering invited talks in several conferences, symposiums, and faculty development programs (FDPs), he has organized several workshops, symposiums, and international conferences. He has been awarded second prize in the academic writing competition organized by SIAM journal publishing. He has more than 40 research articles in journals of high repute, namely, *Scientific Reports* (*Nature*), *Reliability Engineering and System Safety*, *Journal of Computational and Applied Mathematics*, *Results in Engineering*, *Quality Technology & Quantitative Management*, *Arabian Journal for Science and Engineering*, and *Engineering Reports*. He is also a reviewer of many reputed journals. As a professional, he has visited IIRS (ISRO), CSIR-IIP, and WIHG. Additionally, he has completed one research project (Short-Term Empirical Research Project funded by ICSSR, New Delhi) in 2023 and is a part (as a CO-PI) of an ongoing research project of Rs. 1.04 crore funded by ICSSR, New Delhi.

Jyotiranjan Nayak is a professor in the Department of Mathematics at the Faculty of Science and Technology (ICFAI Tech), IFHE, Hyderabad, India. Prior to this, he was an associate professor at the Institute of Technical Education and Research, Siksha 'O' Anusandhan University, Bhubaneswar, India, and has more than 32 years of teaching experience in undergraduate and postgraduate engineering studies. He has taught subjects like differential equations, vector analysis, complex analysis, linear algebra, numerical methods, optimization techniques, discrete mathematics, and advanced calculus at the undergraduate level. He has taught postgraduate students algebra, cryptography, number theory, real analysis, topology, combinatorics, graph theory, and operations research. For PhD students, he has delivered lectures on advanced computational mathematics, matrix algebra, and design of experiments at different universities, including Utkal University, Biju Patnaik University of Technology, Berhampur University, Siksha 'O' Anusandhan, and ICFAI University. Besides, he has been a guest and visiting faculty at many other universities in the country. He obtained his PhD from Utkal University in 2005. His research areas include non-convex programming, fuzzy optimization, optimal control, and topology optimization. He has published 12 research papers in leading national and international journals. He is also a former associate editor of *Journal of Orissa Mathematical Society*. He was the vice-president of the Operational Research Society of India (ORSI), Bhubaneswar Chapter. He is an executive member of the Orissa Mathematical Society (OMS), a life member of the Indian Society for Technical Education (ISTE), Orissa Information Technology Society (OITS), Ramanujan Society of Mathematics and Mathematical Sciences (RSMSS).

Mayank Gupta received his PhD from IIT Delhi. He completed his MSc physics from the Department of Physics & Astrophysics, Delhi University, and graduated with a BSc (Hons) in physics from Hansraj College, Delhi University. Before joining PDEU, he worked as a project scientist at IIT Delhi on a project of Defence Research and Development Organisation (funding of around 7 crores). The main area of his research is optics and solar energy. He has developed products for indoor daylighting, solar thermal, and hybrid energy generation systems without solar tracking. He has published more than 14 international research papers with high-impact factors and attended more than 10 international conferences in India and abroad. He has six patents granted in his name, and two are published. One technology is transferred to the industry for the product developed based on his research work in solar energy. He is also a scientific reviewer of several international journals such as *Optics Express* (OSA), *Optics Letters* (OSA), *Applied Optics* (OSA), MDPI journals, and many others. He is on the editorial board of several international journals. He has been working in collaboration with industries such as

ReNew Power Pvt. Ltd. (Gurugram), Maharishi Solar Thermal Pvt. Ltd. (Noida), EP Sunsol Pvt. Ltd. (Chennai), Industrial Research and Development Unit (IRD, IIT Delhi), and Foundation for Innovation and Technology Transfer (FITT, IIT Delhi).

Chandra Shekhar is a professor, a distinguished academic, and a former head of the Department of Mathematics at BITS Pilani, India. He is actively involved in research and teaching at both undergraduate and postgraduate levels. His expertise spans various mathematical fields, including queueing theory, inventory theory, reliability theory, computer and communication systems, machine repair problems, stochastic processes, evolutionary computation, statistical analysis, and fuzzy set and logic. He actively participates in national and international conferences, seminars, symposiums, workshops, and FDPs, and has organized numerous events as a convener and organizing secretary. His paper received the Best Research Paper Award at an international conference. With over 55 research articles in esteemed journals and book chapters, he has also supervised four PhD theses, authored a textbook entitled *Differential Equations, Calculus of Variations, and Special Functions*, and edited four reference books. Beyond academia, he serves as a member of editorial boards and academic societies, and reviews for prestigious journals. He serves on multiple academic committees, including those for Hon'ble Governor/Chancellor nominees, board of management, doctoral research committees, board of studies, advisory boards, faculty selection committees, and examination boards for government and private universities, institutions, and research laboratories. Additionally, he has collaborated with and visited several renowned organizations, including IIRS (ISRO), CSIR-IIP, NIH, WIHG, CPWD, NTPC, Ordnance Factory, Survey of India, Indian Red Cross Society, Bank of Maharashtra, and APS Lifetech.

Balakrushna Tripathy is working as a professor in the higher academic grade at the CORE school of VIT, Vellore. He received his PhD in 1983. During his student career, he received three gold medals for standing first at the graduation level, standing first at the postgraduate level, and being adjudged as the best postgraduate of the year from Berhampur University, Odisha. He has the distinction of receiving a national scholarship at the postgraduate level, University Grants Commission (government of India) fellowship for pursuing his research, and Department of Science & Technology (government of India) fellowship for pursuing M. Tech. (computer science) at Pune University, and the SERC fellowship (DOE, government of India) for joining IIT Kharagpur as a visiting fellow. Dr. Tripathy is a fellow of IETE and IEI. He is among the 2% scientists in the world for the year 2022, and his career is also counted as measured by Stanford University, USA. He has a distinguished research career in both mathematics and computer science, resulting in the publication of more than 750 articles in international journals, proceedings of international conferences of repute, and chapters in edited research volumes. Also, he has edited eight research volumes and written two books and two monographs. During his career so far, he has supervised candidates for 32 PhD degrees, 13 M. Phil. degrees, and 5 MS and M. Tech by research degrees. Also, he has guided numerous student projects at the B. Tech, M. Tech, and MCA levels. Dr. Tripathy is a member of more than 20 professional bodies, a senior member of IEEE, a senior member of ACM, a senior member of IRSS, and a senior life member of CSI, the Indian Academy of Mathematics, IEEE Communication Society, Indian Mathematical Society, ACM Compute Newsgroup, Indian Science Congress Association, International Science and Technology group, and IEEE Fuzzy Logic Society. In recognition of his distinguished research in the field of mathematics, the American Mathematical Society included him as a reviewer for *Mathematical Reviews* in 1982 and later honored him with an honorary membership of the American Mathematical Society in 1992. In 1983, he was selected as a reviewer for Zentralblatt fur Mathematik (Germany). At present, Dr. Tripathy is a reviewer for over 100 international journals from all top publishing houses in the world. Some of the journals of repute where he has published articles include *Information Sciences*, *Applied Soft Computing*, *IEEE Access*, *Journal of Mathematical Analysis and Applications*, *Bulletin of the Malaysian Mathematical Society*, *Indian Journal of Pure and Applied Mathematics*, *Journal of Educational Technology & Society*, *International Journal of Earth Sciences and Engineering*, *Kybernetes*, *World Applied Sciences Journal*, *International Journal of Communication Systems*, *Journal of Web Engineering*, *Evolutionary Intelligence*, *International Journal of Uncertainty, Fuzziness and Knowledge-Based Systems*, *Frontiers in Public Health*, and *Sensors* (Basel), to name a few. He has evaluated the PhD theses of about 21 universities in India. He has been sanctioned with three research projects from the central agencies in India. Also, Dr. Tripathy has served in several administrative positions like head of the departments, program manager of different academic programs, division leader of research groups, and dean of schools.

Contributors

M. M. Acharya
Department of Mathematics
School of Applied Sciences
KIIT Deemed to Be University
Bhubaneswar, Odisha, India

Srikumar Acharya
Department of Mathematics
School of Applied Sciences
KIIT Deemed to Be University
Bhubaneswar, Odisha, India

Asipu Anusha
Department of Mathematics
KIIT Deemed to Be University
Bhubaneswar, Odisha, India

Aksh Bavisi
Department of Mechanical Engineering
School of Technology
Pandit Deendayal Energy University
Gandhinagar, Gujarat, India

Siva Prasad Behera
Department of Mathematics
C.V. Raman Global University
Bhubaneswar, Odisha, India

Ankit Deshmukh
Department of Civil Engineering
School of Technology
Pandit Deendayal Energy University
Gandhinagar, Gujarat, India

Topunuru Kaladhar
Department of Economics
ICFAI School of Social Sciences
Hyderabad, Telangana, India

Bhuvnesh Khatana
Department of Mathematics
Indian Institute of Technology Kharagpur
Kharagpur, West Bengal, India

P. Ajith Kumar
Department of Finance and Accounting
ICFAI Business School (IBS)
Hyderabad, Telangana, India

Alok Kumar
FORE School of Management
New Delhi, India

A. P. Maharana
Department of Mathematics
Siksha 'O' Anusandhan (Deemed to Be University)
Bhubaneswar, Odisha, India

R. Manjula
School of Computer Science and Engineering
VIT, Vellore, India

Kishor Chandra Meher
Department of Accounting and Finance
University of Fiji, Saweni, Fiji

Ananya Mishra
Department of Mathematics
C.V. Raman Global University
Bhubaneswar, Odisha, India

S. K. Misra
Department of Mathematics
KIIT Deemed to Be University
Bhubaneswar, Odisha, India

N. Munikumar
School of Computer Science and Engineering
VIT, Vellore, India

S. Nanda
Department of Mathematics
School of Applied Sciences
KIIT Deemed to Be University
Bhubaneswar, Odisha, India

Rashmi Ranjan Ota
Department of Mathematics
Institute of Technical Education and Research
SOA (Deemed to Be) University
Bhubaneswar, Odisha, India

Geetanjali Panda
Department of Mathematics
Indian Institute of Technology Kharagpur
Kharagpur, West Bengal, India

Ashutosh Parhi
KIIT School of Management (KSOM) KIIT Deemed to be University
Bhubaneswar, India

Rahul Priyadarshi
Symbiosis Institute of International Business
Symbiosis International University (Deemed to Be University)
Pune, Maharashtra, India

P. K. Rout
KIIT School of Biotechnology
KIIT Deemed to Be University
Bhubaneswar, Odisha, India

Saroj Kumar Routray
School of Management
KIIT Deemed to Be University
Bhubaneswar, Odisha, India

Anuradha Sahoo
Department of Mathematics
Siksha 'O' Anusandhan (Deemed to Be University)
Bhubaneswar, Odisha, India

Manthan Shah
Department of Mechanical Engineering
School of Technology
Pandit Deendayal Energy University
Gandhinagar, Gujarat, India

Chandra Shekhar
Department of Mathematics
Birla Institute of Technology and Science
Pilani Campus, Pilani, Rajasthan, India

Bhasuru Abhinaya Srinivas
Department of Mechanical Engineering
School of Technology
Pandit Deendayal Energy University
Gandhinagar, Gujarat, India

Anurag Kumar Srivastava
Department of Public Policy and Public Administration
School of Liberal Studies (SLS)
Pandit Deendayal Energy University
Gandhinagar, Gujarat, India

Rabi Narayan Subudhi
School of Management
Centurion University of Technology and Management
Sitapur, Odisha, India

Sushree Swetapadma
Department of Mathematics
KIIT Deemed to Be University
Bhubaneswar, Odisha, India

Balakrushna Tripathy
School of Computer Science
Engineering and Information Systems
VIT, Vellore, India

Shreekant Varshney
Department of Mathematics
School of Technology
Pandit Deendayal Energy University
Gandhinagar, Gujarat, India

Amit Yadav
Symbiosis Institute of International Business
Symbiosis International University (Deemed to Be University)
Pune, Maharashtra, India

Chapter 1

Sensitivity Investigation of Finite Queueing Systems with Unreliable Service Provisioning and Feedback Mechanism

Shreekant Varshney, Aksh Bavisi, Manthan Shah, Chandra Shekhar, and Anurag Kumar Srivastava

1.1 INTRODUCTION

The theory of queues offers critical statistical techniques for modeling service systems in which customers arrive, wait, receive service, and depart. The single-server finite queue provides an essential foundational component in this framework, which is defined by Markovian arrival and service operations with a single-service provider. Modern service systems usually operate in non-ideal conditions where server breakdowns occur at random, disrupting service provisioning and affecting system performance. Additionally, feedback mechanisms, where customers may demand several service requests before successful completion, add intricate dynamics that substantially affect operational efficiency. The inclusion of server breakdowns and feedback mechanisms into conventional queueing models reflects real-world operational challenges across diverse applications. Modern service mechanisms include cloud computing infrastructure, manufacturing systems, telecommunication networks, and healthcare delivery systems that often comply with identical characteristics. Server failures represent inherent operational constraints of technological systems, whereas service quality challenges needing feedback from customer channels are frequent across service industries. The correlation between these aspects results in complex system dynamics needing advanced mathematical approaches for accurate performance prediction and optimization.

Service providers must efficiently balance numerous operational concerns: preserving system stability despite server failures, meeting customer feedback requirements, and optimizing resource allocation. Understanding the complex relationships among arrival patterns, service mechanisms, breakdown occurrences, and feedback cycles becomes important for developing effective operational strategies. This analysis proposes a thorough analytical framework incorporating various operational factors while maintaining practical relevance across diverse customer service environments. The flowchart illustrating the working procedure for the queueing infrastructure based on the suggested queueing framework is presented in Figure 1.1.

DOI: 10.1201/9781003642091-1

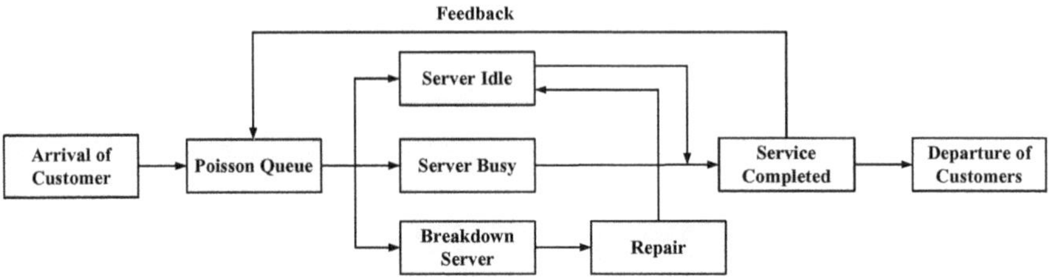

Figure 1.1 Flowchart representing the working procedure of the queueing infrastructure based on the proposed queueing framework.

The evolution of queueing theory has witnessed significant advancements through mathematical innovations and practical implementations. While classical single-server Markovian queues have been extensively investigated, the addition of server failures presents reliability engineering perspectives that expand traditional queueing analysis. Recent research has examined numerous aspects of unreliable servers, continuously exploring breakdown patterns, repair procedures, and system stability. Simultaneously, feedback mechanisms have gained significance in quality-focused service environments, becoming new statistical challenges in performance measurement and service enhancement. Queueing-based stochastic modeling has advanced enormously, with the classical queueing models serving as a core foundation for several analytical advancements. Ramesh and Udayabaskaran [1] examined single-server queueing systems operating in doubly stochastic environments, while Strzęciwilk [2] deployed timed Petri nets for simulating priority queueing systems. Server reliability has attracted substantial attention, with Varshney et al. [3] investigating finite capacity service systems with partial server breakdown and recovery mechanisms. Shekhar et al. [4] explored multi-server finite capacity systems with Bernoulli's scheduled vacation policy. Feedback mechanisms have been thoroughly investigated by Varshney et al. [5], who constructed controlled queueing systems using Markovian modeling. Shah et al. [6] implemented queueing networks to assess various charging stations in electric vehicle (EV) networks. Furthermore, the retrial phenomena have emerged as essential characteristics in modern systems (cf. Hablal et al. [7], Nithya et al. [8], Kumaresan et al. [9], Wang et al. [10], Li et al. [11]), establishing constraints on simultaneous stationary distributions and assessing state-dependent arrivals in queueing-inventory systems.

Ata et al. [15] introduced computational approaches for dealing with independent stochastic control problems, demonstrating how advanced mathematical tools have improved analytical capabilities. Szavits-Nossan and Grima [17] employed the concept of queues to model stochastic individual perspectives. Mehri and Ciucu [16] established continuous-time stochastic analysis for precise closed-form outcomes. Besides, EV infrastructure applications have been thoroughly examined by Varshney et al. [12], who incorporated working vacations and arrival control mechanisms. Varshney et al. [13] established stochastic models for multiple-server charging stations employing feedback methods. Furthermore, renewable energy systems were examined by Varshney et al. [14] utilizing matrix-analytical methods for optimal management strategies. Edwan and Tahat [18] proposed analytical ON-OFF queueing mobility models for software-defined vehicle networks. Customer impatience has attracted extensive attention (cf. Li et al. [19], Bekker et al. [20], Liu et al. [21], Xie et al. [22], Wang et al. [23]), exploring queueing switching patterns, integrated routing solutions, and healthcare applications.

Recent investigation demonstrates substantial advancements in the theory of queues for complex service systems, particularly in addressing feedback mechanisms, unreliable servers, and customer impatience. Shah et al. [6] constructed a Markovian framework for EV charging networks that incorporates abandonment and feedback, whereas Varshney et al. [24] and Nayak et al. [25] focused on optimizing performance in systems with unreliable servers using nature-inspired algorithms. In the recent past, numerous authors (cf. Shekhar et al. [26], Shekhar et al. [27], Shekhar et al. [28], Gupta et al. [29], and Varshney et al. [13]) have comprehensively analyzed single- and multi-server queueing systems with vacation policies, reneging, and feedback mechanisms. Moreover, Varshney et al. [30] employed equivalent queueing modeling to renewable energy and EV infrastructure applications, while Shekhar et al. [31] examined transient solutions and optimal control via meta-heuristic techniques. Collectively, these studies underline the importance of integrated stochastic models and advanced optimization in boosting service quality and economic efficiency in modern service environments (cf. Tables 1.1 and 1.2).

Optimization methodologies have developed enormously, with Varshney et al. [13] applying evolutionary algorithms for charging infrastructure, while Mathavavisakan and Indhira [36] accomplished cost optimization through particle swarm optimization. Samouylov et al. [37] established mathematical frameworks for managing correlated request flows through priority-based extraction rates. Nayak et al. [25] addressed timetabling as a challenging topic for investigation with applications in education and transportation. Roy et al. [38] examined transient solutions for customer discouragement and system dynamics. Singh et al. [39] explored several features of queueing systems, including abandonment and feedback. Hasan et al. [40] and Wang [41] enhanced web service performance modeling, while Kushvaha et al. [42] examined service distributions and hybrid priority systems. Despite these developments, research gaps persist in frameworks that collectively deal with server reliability challenges and feedback dynamics, notably in transient behavior analysis following breakdown situations. The economic effects of server failures and feedback processes are inadequately assessed, indicating the prospects of developing integrated frameworks integrating

Table 1.1 Recent Studies on Queueing Modeling with Multiple Features and Optimization Approaches.

Authors	*Methodology*	*Key Findings*
Varshney et al. [12]	Markovian modeling, Laplace transformation	Effective management of customer arrival and departure through explicit control strategies
Varshney et al. [13]	Queueing-based stochastic modeling, evolutionary algorithms	Optimal charging infrastructure through comparative investigation of algorithms
Varshney et al. [14]	Matrix-analytical method, Cuckoo search	Optimal expected cost of renewable energy systems with hyper-exponential service
Hablal et al. [7]	Stochastic comparison methods	Lower and upper bounds for joint stationary distributions in priority retrial systems
Ata et al. [15]	Computational stochastic control	Nearly optimal solutions for high-dimensional queueing systems
Mehri and Ciucu [16]	Continuous-time martingale analysis	Exact results in closed form for complex queueing systems with batch arrivals
Shekhar et al. [4]	Matrix-geometric approach, meta-heuristic optimization	Cost minimization is achieved through the retention policy of reneged customers

Table 1.2 Applications of Queueing-Based Stochastic Modeling in Various Domains with Specific Characteristics.

Authors	*Application Domain*	*Major Contributions*
Shah et al. [6]	EV networks	Effective solutions for charging infrastructure considering queueing management
Varshney et al. [4]	Renewable energy systems	Matrix-analytical methods for performance optimization in energy generation
Liu et al. [21]	Healthcare services	Comprehensive approaches for patient flow optimization
Zin et al. [32]	Smart city infrastructure	Integration of queueing theory with artificial intelligence (AI) methodologies
Li [33]	Cloud computing	Resource allocation models accounting for dynamic workload characteristics
Saini et al. [34]	Manufacturing systems	Enhanced efficiency through analysis of arrival rates and system capacities
Anagha and Sreelatha [35]	Supermarket operations	Practical insights for staffing and resource allocation during peak hours

reliability engineering with feedback-oriented service quality considerations in queueing frameworks.

Based on an extensive investigation of the existing literature, several critical research gaps have been identified:

- Insufficient incorporation of multiple system dynamics: While server failures and feedback mechanisms have been researched independently, insufficient attention has been given to their combined consequences within finite queueing systems.
- Inadequate analytical frameworks: Existing mathematical models often fail to capture the complex interrelationships among breakdown occurrences, repair processes, and feedback probabilities in unified formulations.
- Restricted performance and sensitivity analyses: Previous studies have primarily focused on basic performance metrics without exploring the comprehensive impact of breakdown parameters and feedback probabilities on system behavior.
- Insufficient optimization approaches: Limited research exists on developing cost-effective operational strategies that simultaneously address server reliability challenges and feedback management.
- Lack of dynamical investigations: Most existing studies focus exclusively on steady-state analysis, overlooking the critical transient/dynamical behavior during system recovery after breakdowns.
- Inadequate research of economic implications: The economic implications of server breakdowns and feedback mechanisms have not been effectively assessed through detailed cost modeling.

Furthermore, the primary objectives and novel contributions of the present investigation are illustrated as follows:

- To develop a unified mathematical framework for analyzing single-server finite queueing systems incorporating both server breakdown phenomena and customer feedback mechanisms.
- To establish closed-form expressions for steady-state key performance measures under varying breakdown circumstances and feedback probabilities.
- To derive analytical relationships between system parameters (arrival rate, service rate, breakdown rate, repair rate, and feedback probability) and performance metrics.
- To provide practical insights for system designers and operators through extensive numerical investigations demonstrating parameter sensitivity and performance trade-offs.
- To propose adaptive control strategies that dynamically adjust system parameters in response to changing breakdown patterns and feedback requirements.

This research bridges theoretical foundations with practical implementations in modern service environments where server reliability and service quality simultaneously impact system performance and operational economics.

The present study's general framework is outlined as follows: Section 1.2 describes the steady development of stochastic modeling through multiple hypotheses and annotations. The governing set of Chapman–Kolmogorov differential equations is constructed by applying the fundamental concepts of birth and death phenomena and Markov chains that are presented in subsection 1.2.1. Section 1.3 emphasizes an elementary methodological approach for obtaining the state probability distribution of the developed stochastic model through the integration of matrix solution approaches. Section 1.4 provides closed-form equations for numerous system performance characteristics to analyze the overall effectiveness of queueing networks. An in-depth parametric and sensitivity investigation is carried out utilizing numerous significant examples that validate the suggested modeling and investigation's conclusions, as indicated by several graphs and tables in Section 1.5. Section 1.6 emphasizes various crucial managerial recommendations that could benefit system designers, decision-makers, researchers, and scholars in the process of decision-making. Lastly, Section 1.7 offers concluding perspectives, numerous potential future directions, and an outline of prospective extensions of the suggested queueing-based stochastic modeling.

1.2 MODEL ASSUMPTIONS AND NOTATIONS

This study considers a finite capacity queueing system that integrates server breakdown phenomena with a customer feedback mechanism. Customers arrive following a Poisson process with rate λ and are served on a first-come-first-served basis by a single server whose service time is independently and exponentially distributed with rate parameter μ. The system can accommodate up to K independent customers, and the server is subject to experiencing random breakdown occurrences with rate α, after which it is repaired immediately with rate parameter β. Additionally, upon service completion, customers may require additional service due to unsatisfactory service provisioning, re-entering the system according to a feedback mechanism characterized by probability p, otherwise, leave permanently the queueing system with the complementary probability $q = (1 - p)$. These assumptions and notations: K for system capacity; λ for the arrival rate; μ for the service rate; α and β for the breakdown and repair rates, respectively; and p for the feedback probability, provide the foundation for analyzing system performance under ideal operational conditions.

Now, let us suppose that $J(t)$ is the state of the service provider at time instant t and $N(t)$ is the number of customers in the queueing system at time instant t. Thus, the process $\{(J(t), N(t)), t \geq 0\}$ constitutes an

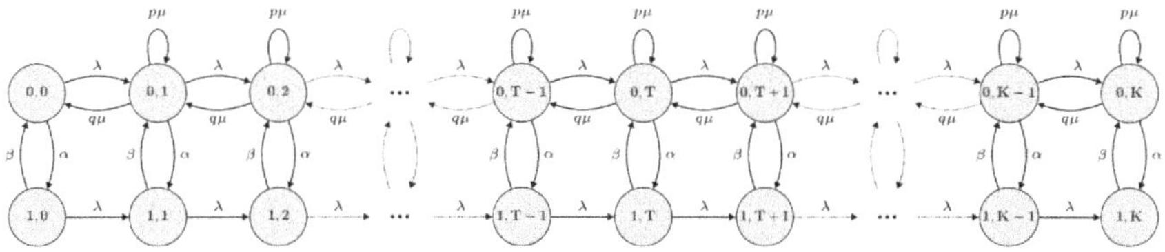

Figure 1.2 Transition state diagram associated with the proposed queueing framework.

irreducible continuous-time Markov chain (CTMC) with the state-space $\Omega = \{(j,n),\ j = 0, 1, \text{ and } n = 0, 1,\ldots,K\}$. Furthermore, combining the input and output flow from system states and employing the notion of CTMC, the state transition diagram (cf. Figure 1.2) of the developed model can be illustrated as

1.2.1 Differential-Difference Equations

In queueing simulation, differential-difference equations become crucial because they provide an environment for evaluating dynamic systems in which ongoing operations and specific incidents interconnect with one another. These equations allow scholars and investigators to quantify the novel features of customer arrivals, service mechanisms, and system dynamics across varying time periods, making them crucial for understanding the behavior of complex queueing systems. By incorporating both differential and difference characteristics, these equations improve the modeling of transient states and time-dependent events, thereby boosting the accuracy of performance predictions such as waiting times, queue lengths, and overall throughput of the service systems. Consequently, their application becomes essential for boosting service efficiency and optimizing decision-making in varied operating configurations. Furthermore, the differential-difference equations correlated with the established queueing framework with the unreliable server, and the feedback mechanism can be defined as follows:

$$\frac{dP_{0,0}(t)}{dt} = -(\lambda + \alpha)P_{0,0}(t) + q\mu P_{0,1}(t) + \beta P_{1,0}(t), \tag{1.1}$$

$$\begin{aligned}\frac{dP_{0,n}(t)}{dt} &= -(\lambda + \alpha + q\mu)P_{0,n}(t) + \lambda P_{0,n-1}(t) + \beta P_{1,n}(t) + q\mu P_{0,n+1}(t),\\ &\quad 1 \le n \le K-1,\end{aligned} \tag{1.2}$$

$$\frac{dP_{0,K}(t)}{dt} = -(\alpha + q\mu)P_{0,K}(t) + \lambda P_{0,K-1}(t) + \beta P_{1,K}(t), \tag{1.3}$$

$$\frac{dP_{1,0}(t)}{dt} = -(\lambda+\beta)P_{1,0}(t) + \alpha P_{0,0}(t), \tag{1.4}$$

$$\frac{dP_{1,n}(t)}{dt} = -(\lambda+\beta)P_{1,n}(t) + \lambda P_{1,n-1}(t) + \alpha P_{0,n}(t), 1 \le n \le K-1, \tag{1.5}$$

$$\frac{dP_{1,K}(t)}{dt} = -\beta P_{1,K}(t) + \lambda P_{1,K-1}(t) + \alpha P_{0,K}(t). \tag{1.6}$$

1.2.2 Steady-State Equations

In the equilibrium, i.e., as $t \to \infty$, the aforementioned system of differential equations (1.1)–(1.6) can be converted to

$$(\lambda+\alpha)P_{0,0} = q\mu P_{0,1} + \beta P_{1,0}, \tag{1.7}$$

$$(\lambda+\alpha+q\mu)P_{0,n} = \lambda P_{0,n-1} + \beta P_{1,n} + q\mu P_{0,n+1}, \quad 1 \le n \le K-1, \tag{1.8}$$

$$(\alpha+q\mu)P_{0,K} = \lambda P_{0,K-1} + \beta P_{1,K}, \tag{1.9}$$

$$(\lambda+\beta)P_{1,0} = \alpha P_{0,0}, \tag{1.10}$$

$$(\lambda+\beta)P_{1,n} = \lambda P_{1,n-1} + \alpha P_{0,n}, \quad 1 \le n \le K-1, \tag{1.11}$$

$$\beta P_{1,K} = \lambda P_{1,K-1} + \alpha P_{0,K}. \tag{1.12}$$

Additionally, together with these steady-state equations, the normalizing condition can also be explicitly written as

$$\sum_{j=0}^{1}\sum_{n=0}^{K} P_{j,n} = 1 \tag{1.13}$$

1.3 MATRIX SOLUTION METHODOLOGY

Matrix-analytical technique is a powerful methodology for identifying and developing stochastic models. This technique is commonly applied in dealing with varied queueing situations that require appropriate numerical interpretations. By applying the matrix technique, the abovementioned set of differential equations (1.7)–(1.12) can be transformed into the matrix expression $\pi Q = 0$, where π is the steady-state probability vector of dimension $(2K+2)$ and Q is the generator matrix corresponding to all the state probabilities. The structure of the correlated block tri-diagonal rate matrix Q correlated with the underlying Markov chain and having dimension $(2K+2)\times(2K+2)$ can be illustrated as

$$
Q = \begin{bmatrix}
A_0 & B & 0 & \cdots & 0 & 0 & 0 & \cdots & 0 & 0 & 0 \\
C & A_1 & B & \cdots & 0 & 0 & 0 & \cdots & 0 & 0 & 0 \\
0 & C & A_1 & \cdots & 0 & 0 & 0 & \cdots & 0 & 0 & 0 \\
\vdots & \vdots & \vdots & \ddots & \vdots & \vdots & \vdots & \vdots & \vdots & \vdots & \vdots \\
0 & 0 & 0 & \cdots & A_1 & B & 0 & \cdots & 0 & 0 & 0 \\
0 & 0 & 0 & \cdots & C & A_1 & B & \cdots & 0 & 0 & 0 \\
0 & 0 & 0 & \cdots & 0 & C & A_1 & \cdots & 0 & 0 & 0 \\
\vdots & \vdots & \vdots & \cdots & \vdots & \vdots & \vdots & \ddots & \vdots & \vdots & \vdots \\
0 & 0 & 0 & \cdots & 0 & 0 & 0 & \cdots & A_1 & B & 0 \\
0 & 0 & 0 & \cdots & 0 & 0 & 0 & \cdots & C & A_1 & B \\
0 & 0 & 0 & \cdots & 0 & 0 & 0 & \cdots & 0 & C & A_2
\end{bmatrix},
$$

where

$$
A_0 = \begin{bmatrix} -(\lambda+\alpha) & \alpha \\ \beta & -(\lambda+\beta) \end{bmatrix}, \quad B = \begin{bmatrix} \lambda & 0 \\ 0 & \lambda \end{bmatrix},
$$

$$
A_1 = \begin{bmatrix} -(\lambda+\alpha+q\mu) & \alpha \\ \beta & -(\lambda+\beta) \end{bmatrix},
$$

$$
A_2 = \begin{bmatrix} -(\alpha+q\mu) & \alpha \\ \beta & -\beta \end{bmatrix} \text{and } C = \begin{bmatrix} q\mu & 0 \\ 0 & 0 \end{bmatrix}.
$$

Here, the state probability vector associated with the rate matrix Q in the equilibrium is represented by π, which can be partitioned into $\pi = [\pi_0, \pi_1, \pi_2, \cdots, \pi_{T-1}, \pi_T, \pi_{T+1}, \cdots, \pi_{K-1}, \pi_K]$, where the sub-vectors $\pi_n = [P_{0,n}, P_{1,n}]$ for $n = 0,1,2,\ldots,K$ are of dimension (1×2). Let us suppose that vector **e** represents a column vector of dimension $(2\ K+2)$, all of which elements as ones. Now, utilizing and solving the homogeneous system of linear equations $\pi\mathbf{Q} = 0$ and employing the normalizing condition $\pi\mathbf{e} = 1$, the steady-state probabilities associated with the abovementioned matrix equations can be easily demonstrated numerically with the help of MATLAB (2024b) software.

1.4 SYSTEM PERFORMANCE MEASURES

In queueing simulation, the necessity for different system performance metrics plays an important role in creating effective management strategies and enhancing service efficiency. The expected number of consumers in the system, the probability that the server is in the breakdown state, the system's overall throughput, server utilization, and many other crucial metrics provide substantial details about how the system operates and the manner in which it performs. By examining such factors, professionals and scholars might recognize challenges, optimize the distribution of resources, and develop appropriate system operations that satisfy fluctuating demand patterns. Furthermore, integrating numerous performance indicators provides for a more comprehensive understanding of possibilities throughout the system, enabling precise decision-making that eventually leads to improved customer experiences and operational effectiveness in varied service conditions.

1) Expected number of customers in the system:

$$L_S = \sum_{j=0}^{1}\sum_{n=0}^{K} nP_{J,n}. \tag{1.14}$$

2) Expected number of customers in the queue:

$$L_Q = \sum_{n=1}^{K} nP_{0,n} + \sum_{n=0}^{K} nP_{1,n}. \tag{1.15}$$

3) Probability that the server is in the breakdown state:

$$P_B = \sum_{n=0}^{K} P_{0,n}. \tag{1.16}$$

4) Probability that an immediate repair is provided to the breakdown server:

$$P_R = \sum_{n=0}^{K} P_{1,n}. \tag{1.17}$$

5) Average rate of individuals rejoining the system to complete their unsatisfactory service:

$$AFM = \sum_{n=1}^{K} (p\mu)P_{0,n}. \tag{1.18}$$

6) Overall throughput of the system:

$$\tau_p = \sum_{n=1}^{K} \mu P_{0,n}. \tag{1.19}$$

7) Expected waiting time of customers in the system:

$$W_S = \left(\frac{L_S}{\lambda}\right). \quad (1.20)$$

1.5 NUMERICAL INVESTIGATION

In this section, we present numerical illustrations to validate the proposed queueing model with server breakdown and feedback mechanisms. These results demonstrate the system behavior under various operational circumstances and provide insights into the complex interrelationships among breakdown rates, repair times, feedback probabilities, and other critical performance metrics. For illustrative purposes, we initialize the system parameter values as $K = 15$, $\lambda = 1.5$, $\mu = 2.5$, $\alpha = 0.1$, $\beta = 3.0$, $p = 0.1$, and $q = 0.9$. The numerical results are presented through various graphical illustrations, enabling comprehensive analysis of parameter sensitivity and system performance trade-offs.

Figure 1.3 illustrates how the arrival rate (λ) affects the expected number of customers (EN) under varying parameters. One can observe from these subfigures that the system capacity (K) shows minimal impact at lower arrival rates but becomes significant at $\lambda > 1.5$, with $K = 20$ reaching $EN \approx 5.7$ at $\lambda = 1.9$ compared to $K = 12$ with $EN \approx 4.2$. Server breakdown rate (α) consistently impacts system congestion, with $\alpha = 0.5$ resulting in 49% higher EN values than $\alpha = 0.1$ at $\lambda = 1.9$. Repair rates (β) demonstrate modest effects, with faster repairs $(\beta = 4.0)$ reducing EN by approximately 8.5% compared to slower repairs $(\beta = 2.0)$ at high arrival rates. Feedback probability (p) emerges as the most influential parameter, with the lowest value of the complement of the feedback probabilities $(q = 0.1, 0.2)$ maintaining stable $EN \approx 14 - 15$ regardless of the arrival rate, while higher probabilities $(q = 0.4, 0.5)$ show significant convergence as λ increases.

Figure 1.4 demonstrates the impact of service rate (μ) on EN across different system parameter values. As seen in Figure 1.4(i), the system capacity (K) shows a moderate effect at lower service rates, with $K = 20$ yielding $EN \approx 4.2$ at $\mu = 2.1$ versus $K = 12$ with $EN \approx 3.6$, but such difference minimizes as μ increases. Server breakdown rate (α) exhibits a substantial influence throughout, where $\alpha = 0.5$ results in approximately 67% higher EN values $(EN \approx 6.0)$ compared to $\alpha = 0.1$ $(EN \approx 3.6)$ at $\mu = 2.1$ as reflected by Figure 1.4(ii). The repair rate (β) displays an inverse relationship with congestion, where $\beta = 4.0$ reduces EN by roughly 12% compared to $\beta = 2.0$ at $\mu = 2.1$, although this effect becomes negligible at higher service rates as demonstrated by Figure 1.4(iii). Similarly, Figure 1.4(iv) shows that feedback probability (p) emerges as the most critical parameter, with higher values $(q = 0.5)$ showing a significant reduction in EN from 12.8 to 9.2 as μ increases, while lower values $(q = 0.1, 0.2)$

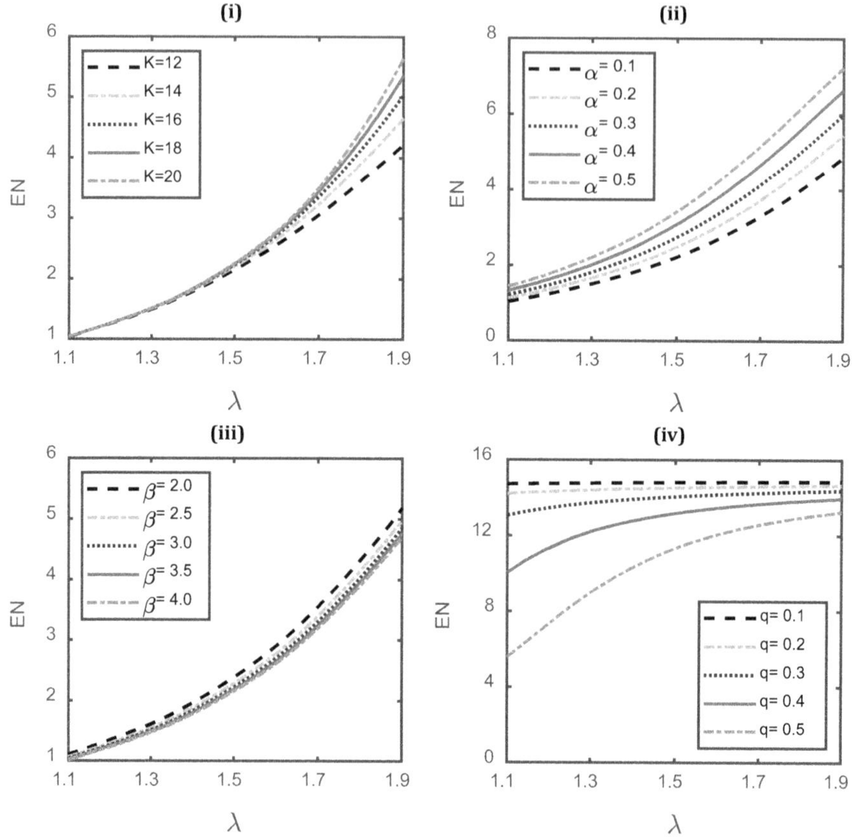

Figure 1.3 Impact of arrival rate (λ) and other system parameters on the expected number of customers in the system.

maintain relatively stable EN levels around 15 regardless of the service rate. These findings align with practical operation management principles, where service improvements yield diminishing returns as other system constraints become limiting factors.

In addition, Figure 1.5 illustrates the correlation between arrival rate (λ) and system throughput (τ_p) under various system parameter values. Figure 1.5(i) reveals that system capacity (K) has minimal impact on throughput across varied values of arrival rates, with output values increasing linearly from $\tau_p \approx 1.1$ at $\lambda = 1.1$ to $\tau_p \approx 1.9$ at $\lambda = 1.9$, showing nearly identical trends regardless of the capacity of the system. Figure 1.5(ii) demonstrates that server breakdown rate (α) similarly exerts negligible influence on throughput, with all configurations following almost identical upward trajectories as λ increases. Figure 1.5(iii) indicates repair rate (β) has a limited effect on system performance, with all values from $\beta = 2.0$ to $\beta = 4.0$

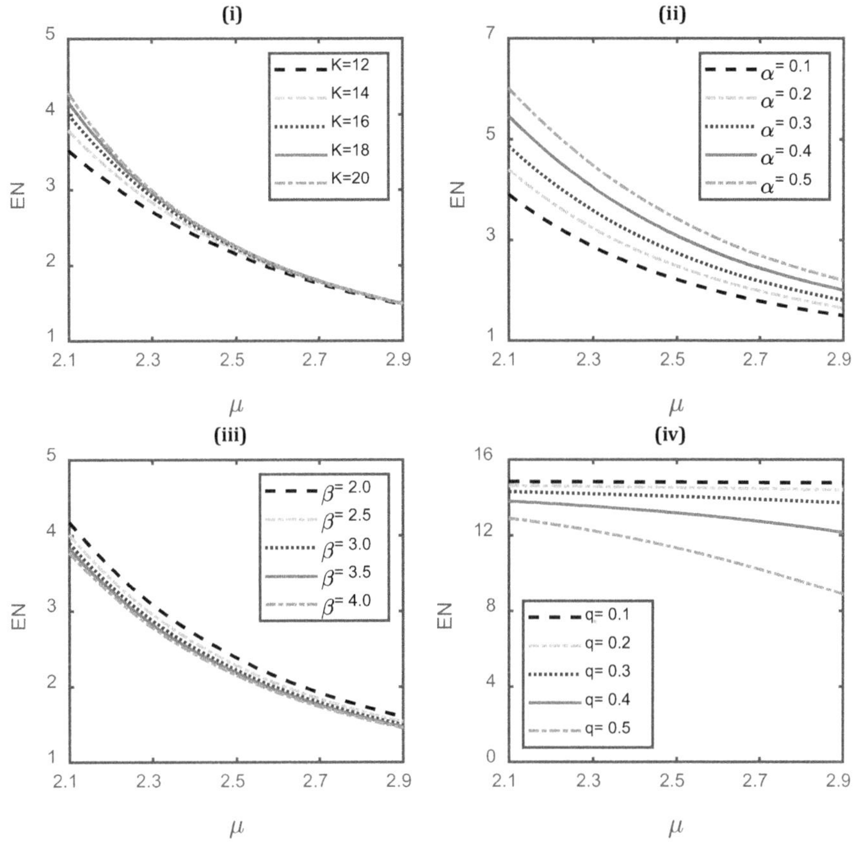

Figure 1.4 Impact of service rate (μ) and other system parameters on *EN* in the system.

producing comparable throughput that increases proportionally with the arrival rate. Most notably, Figure 1.5(iv) highlights feedback probability (p) as the dominant factor affecting throughput, where higher values of the complementary feedback probability $(q = 0.5)$ consistently produce significantly greater τ_p (≈ 1.2) compared to lower values $(q = 0.1, \tau_p \approx 0.2)$ across all arrival rates, with the relative differences maintained throughout. These findings correspond to real-world queueing systems, where customer feedback loops substantially enhance overall system throughput compared to simple changes in system capacity or service reliability parameters.

In a similar manner, Figure 1.6 examines the effect of service rate (μ) on system throughput (τ_p) across different system parameters. Figure 1.6(i) shows that system capacity (K) has minimal influence on throughput, with all configurations exhibiting a slight increase from $\tau_p \approx 1.46$ to $\tau_p \approx 1.49$ as μ rises from 2.1 to 2.9, maintaining consistently small differences between

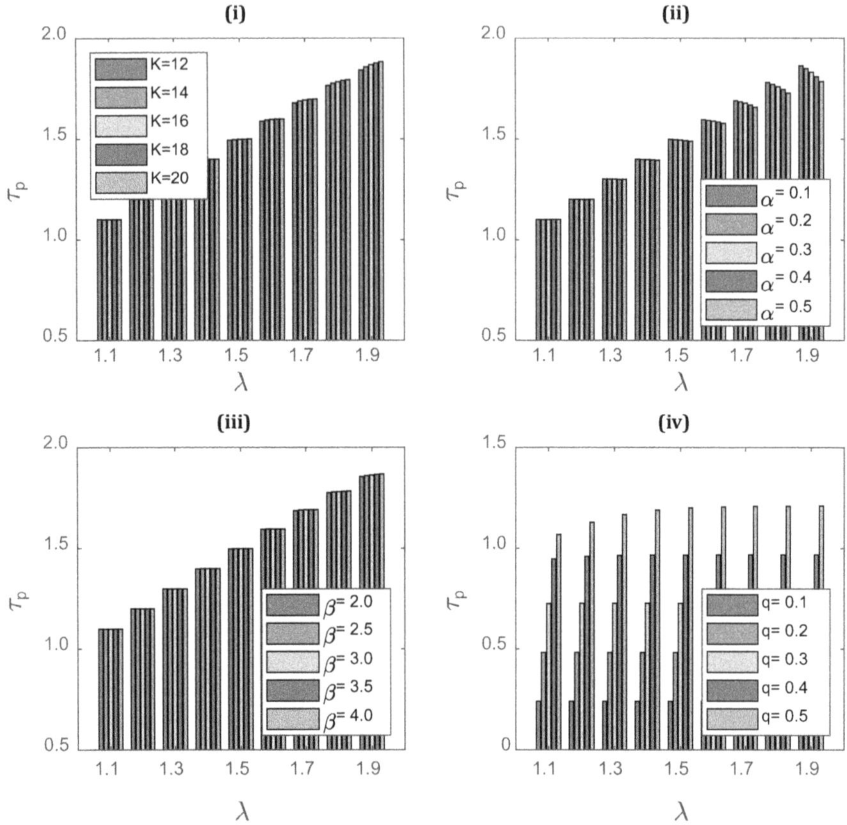

Figure 1.5 Variation in the overall throughput of the system with regard to arrival rate (λ) and other system parameters.

capacity values throughout. Figure 1.6(ii) demonstrates that server breakdown rate (α) similarly produces negligible variations in throughput, with all α values following nearly identical upward trajectories as the service rate increases. Figure 1.6(iii) indicates that repair rate (β) exerts minimal impact on system performance, with the throughput marginally increasing from $\tau_p \approx 1.47$ to $\tau_p \approx 1.49$ across all β configurations as μ increases. In stark contrast, Figure 1.6(iv) reveals feedback probability (p) as the critical determinant of throughput, with higher values of the complementary feedback probability ($q = 0.5$) generating substantially greater τ_p (ranging from 1.0 to 1.4) compared to lower values, $(q = 0.1, \tau_p \approx 0.2)$, and exhibiting more pronounced increases with increasing service rates. These patterns align with theoretical queueing principles, where feedback loops fundamentally enhance system throughput while traditional capacity and reliability parameters yield only incremental performance improvements.

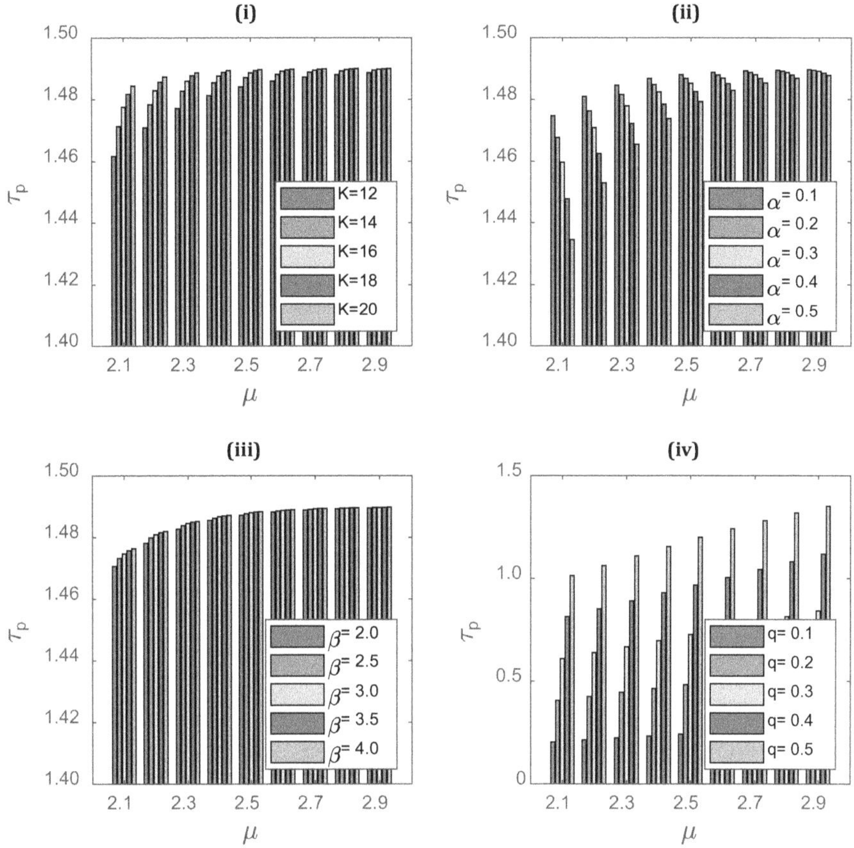

Figure 1.6 Variation in the overall throughput of the system with regard to service rate (μ) and other system parameters.

1.6 SIGNIFICANT MANAGERIAL INSIGHTS

The quantitative outcomes obtained from the current research provide advantageous practical insights for system designers, operations managers, and decision-makers responsible for service systems subjected to server breakdowns and feedback mechanisms. Our sensitivity analysis reveals that investing in server reliability through preventive maintenance significantly reduces the need for reactive maintenance strategies in terms of long-term operating costs and system performance. Managers should establish significant breakdown probability criteria to guide maintenance scheduling decisions, particularly for systems functioning approaching their capacity, when modest improvements in server reliability can considerably enhance key performance indicators. The model demonstrates that minimizing feedback probability by service quality enhancements gives better economic advantages than constructing resources to accommodate high feedback rates. Implementing quality validation checkpoints before service completion can efficiently uncover potential errors during the first service phase, hence lowering overall feedback occurrences.

The optimization results provide straightforward recommendations for resource allocation between reliability improvements and service rate enhancements, enabling evidence-based deployment of limited resources for optimal operational benefits. Systems with higher breakdown rates particularly benefit from prioritizing reliability enhancements over service rate improvements. The findings justify employing threshold-based control policies that dynamically customize acceptance parameters depending on current system circumstances, thereby reducing system overload during periods of increased breakdown probability while preserving service quality standards. The mathematical and statistical modeling approach proposed in this research provides a rigorous framework for capacity planning decisions under uncertainty, enabling the construction of robust service designs that permit operational variability while preventing excess capacity expenditures. By describing straightforward relationships between breakdown rates, feedback probabilities, and performance measures, this approach helps system designers develop balanced operating environments that efficiently manage both reliability concerns and customer feedback requirements.

1.7 CONCLUSION AND FUTURE PERSPECTIVE

This research presents a comprehensive mathematical analysis of the finite capacity queueing systems, incorporating both server breakdown phenomena and customer feedback mechanisms. By developing differential-difference equations and employing the matrix-analytical method, the investigation provides closed-form expressions for critical performance metrics in steady state under varying operational conditions. The parametric and sensitivity analyses reveal critical relationships between breakdown rates, repair times, feedback probabilities, and system performance, highlighting parameter thresholds that significantly impact operational efficiency. The integrated modeling approach efficiently bridges theoretical foundations with practical applications, enabling data-driven operational strategies in environments where server unreliability and customer feedback significantly influence system performance. By quantifying the economic implications of reliability investments and feedback management, the current research provides decision-makers with robust analytical tools for infrastructure planning, resource allocation, and service quality optimization in complex service environments.

Future research directions include extending the framework to accommodate multiple homogeneous and heterogeneous servers with varying breakdown characteristics and incorporating non-Markovian processes with general service time distributions. Also, integration with machine learning approaches could enable real-time prediction of breakdown rates and feedback probabilities, supporting proactive system management. In addition, incorporating more advanced feedback structures with multiple feedback patterns, various server vacation strategies, and retrial phenomena would enhance the model's applicability in complex service environments. These extensions would further advance both theoretical understanding and practical implementation of queueing models in systems characterized by unreliability and feedback requirements.

REFERENCES

1. A. Ramesh, and S. Udayabaskaran, "Analysis of a Single Server Queueing System Controlled by a Random Switch," *Contemporary Mathematics*, vol. 4, no. 2, pp. 189–201, 2023, https://doi.org/10.37256/cm.4220232123.
2. D. Strzęciwilk, "Timed Petri Nets for Modeling and Performance Evaluation of a Priority Queueing System," *Energies (Basel)*, p. 7690, 2023, https://doi.org/10.3390/en16237690.
3. S. Varshney, S. Kaswan, M. Devanda, and C. Shekhar, "Finite Capacity Service System with Partial Server Breakdown and Recovery Policy: An Economic Perspective," *J Syst Sci Syst Eng*, vol. 33, no. 6, pp. 1–31, 2024.
4. C. Shekhar, S. Varshney, and A. Kumar, "Matrix-Geometric Solution of Multi-server Queueing Systems with Bernoulli Scheduled Modified Vacation and Retention of Reneged Customers: A Meta-Heuristic Approach," *Qual Technol Quant Manag*, vol. 18, no. 1, pp. 39–66, 2021.
5. S. Varshney, M. M. Acharya, Z. J. Kanjibhai, M. Pushkarna, W. El-Shafai, and I. Zaitsev, "Economic and Sensitivity Analyses of Controllable Queues with Customer Feedback and Impatience Using Markovian Modeling," *Journal of Mathematics*, vol. 2025, no. 1, Jan. 2025, https://doi.org/10.1155/jom/9648114.
6. M. Shah, S. Varshney, B. A. Srinivas, and A. Bavisi, "Queueing Theoretical Approach and Markov Modeling Employed in Multiple Charging Stations of Electric Vehicle Networks," in *2024 International Conference on Sustainable Energy: Energy Transition and Net-Zero Climate Future (ICUE)*, IEEE, pp. 1–5, Oct. 2024, https://doi.org/10.1109/ICUE63019.2024.10795574.
7. H. Hablal, N. Touche, L. Alem, A. Bouchentouf, and M. Boualem, "Lower and Upper Stochastic Bounds for the Joint Stationary Distribution of a Non-preemptive Priority Retrial Queueing System," *Hacettepe Journal of Mathematics and Statistics*, vol. 52, no. 5, pp. 1438–1460, 2023, https://doi.org/10.15672/hujms.1183966.
8. M. Nithya et al., "Analysis of Stochastic State-Dependent Arrivals in a Queueing-Inventory System with Multiple Server Vacation and Retrial Facility," *Mathematics*, vol. 10, no. 17, p. 3041, 2022, https://doi.org/10.3390/math10173041.
9. V. Kumaresan, J. Bavadharshini, R. Kumar, P. Gokulraj, and G. Hariharan, "Priority-Based Multi-Server Queueing Model for Electric Vehicle Charging," in *2023 4th International Conference on Smart Electronics and Communication (ICOSEC)*, pp. 74–80, 2023, https://doi.org/10.1109/ICOSEC58147.2023.10275815.

10. Q. Wang, D. Zhang, and B. Du, "A Queue Balancing Approach for Electric Vehicle Charging Allocation," in *2022 IEEE 25th International Conference on Intelligent Transportation Systems (ITSC)*, pp. 2750–2755, 2022, https://doi.org/10.1109/ITSC55140.2022.9922375.
11. Z. Li, M. Harchol-Balter, and A. Scheller-Wolf, "Hybrid Priority Queue and Its Applications," *ACM SIGMETRICS Performance Evaluation Review*, pp. 12–14, 2024, https://doi.org/10.1145/3695411.3695417.
12. S. Varshney et al., "Stochastic Modeling of Electric Vehicle Infrastructure Using Queueing-Theoretical Approach," *Results in Engineering*, vol. 25, p. 104149, Mar. 2025, https://doi.org/10.1016/j.rineng.2025.104149.
13. S. Varshney, M. Shah, B. A. Srinivas, M. Gupta, and K. P. Panda, "Queueing Modeling and Optimization of Hybrid Electric Vehicle Infrastructures Using Evolutionary Algorithms," *International Journal of Thermofluids*, vol. 26, p. 101043, 2025, https://doi.org/10.1016/j.ijft.2024.101043.
14. S. Varshney et al., "Optimal Management Strategies of Renewable Energy Systems with Hyperexponential Service Provisioning: An Economic Investigation," *Front Energy Res*, vol. 11, p. 1329899, 2023.
15. B. Ata, J. Harrison, and N. Si, "Singular Control of (Reflected) Brownian Motion: A Computational Method Suitable for Queueing Applications," *Queueing Systems*, vol. 108, no. 3, pp. 215–251.
16. S. Mehri, and F. Ciucu, "On a Continuous-Time Martingale and Two Applications," in *Proceedings of the Twenty-fourth International Symposium on Theory, Algorithmic Foundations, and Protocol Design for Mobile Networks and Mobile Computing*, pp. 111–120, 2023, https://doi.org/10.1145/3565287.3610275.
17. J. Szavits-Nossan, and R. Grima, "Charting the Landscape of Stochastic Gene Expression Models Using Queueing Theory," p. 03253, 2023, [Online]. Available: https://consensus.app/papers/charting-the-landscape-of-stochastic-gene-expression-szavits-nossan-grima/bcef4d1d8c915666affdb4e81e3cf537/
18. T. Edwan, A. Tahat, H. Yanikomeroglu, and J. Crowcroft, "An Analysis of a Stochastic ON-OFF Queueing Mobility Model for Software-Defined Vehicle Networks," *IEEE Trans Mob Comput*, vol. 21, pp. 1552–1565, 2022, https://doi.org/10.1109/tmc.2020.3031319.
19. Y. Li, Z. Cui, J. Ji, and J. Wei, "When do People Switch Queues? An Empirical Study of Discretionary Queue Switching at a Physical Examination Center," *Decision Sciences*, vol. 56, no. 4, pp. 361–382, https://doi.org/10.1111/deci.12646.
20. R. Bekker, B. Bharti, L. Lan, and M. Mandjes, "A Queueing-Based Approach for Integrated Routing and Appointment Scheduling," *Eur. J. Oper. Res.*, vol. 318, pp. 534–548, 2023, https://doi.org/10.1016/j.ejor.2024.05.038.
21. Y. Liu, L. Liu, T. Jiang, and X. Chai, "Strategic Queueing Behavior of Two Groups of Patients in a Healthcare System," *Mathematics*, vol. 12, no. 10, p. 1579, 2024, https://doi.org/10.3390/math12101579.
22. S. Xie et al., "Enhancing Mass Vaccination Programs with Queueing Theory and Spatial Optimization," *Frontiers in Public Health*, vol. 12, p. 1440673.
23. X. Wang, H. Xie, and J. C. S. Lui, "Analyzing Queueing Problems via Bandits with Linear Reward & Nonlinear Workload Fairness," *IEEE Trans Mob Comput*, vol. 23, pp. 3410–3423, 2024, https://doi.org/10.1109/TMC.2023.3276158.
24. S. Varshney, C. Shekhar, V. Tiwari, and K. S. Pritam, "Different Variants of Unreliable Server: An Economic Approach," in *Modeling and Applications in Operations Research*, CRC Press, Baco Raton, 2024, pp. 61–90.

25. J. Nayak, S. Varshney, and C. Shekhar, *Modeling and Applications in Operations Research*. CRC Press, Baco Raton, 2023.
26. C. Shekhar, S. Varshney, and A. Kumar, "Standbys Provisioning in Machine Repair Problem with Unreliable Service and Vacation Interruption," in *The Handbook of Reliability, Maintenance, and System Safety Through Mathematical Modeling*, Elsevier, 2021, pp. 101–133.
27. C. Shekhar, A. Kumar, and S. Varshney, "Load Sharing Redundant Repairable Systems with Switching and Reboot Delay," *Reliab Eng Syst Saf*, vol. 193, Jan. 2020, https://doi.org/10.1016/j.ress.2019.106656.
28. C. Shekhar, A. Kumar, and S. Varshney, "Modified Bessel Series Solution of the Single Server Queueing Model with Feedback," *International Journal of Computing Science and Mathematics*, vol. 10, no. 3, pp. 313–326, 2019.
29. M. Gupta, P. Zala, S. Gupta, and S. Varshney, "Solar Concentration-Based Indoor Daylighting System to Achieve Net Zero Sustainable Buildings," *Energy Build*, vol. 321, 2024, https://doi.org/10.1016/j.enbuild.2024.114662.
30. S. Varshney, B. A. Srinivas, M. Gupta, M. Shah, and A. Bavisi, "Stochastic Modeling of Multiple-Server Charging Stations for Electric Vehicle Networks Using Feedback Strategies: A Queueing-Theoretic Approach," *International Journal of Thermofluids*, vol. 24, p. 100859, 2024, https://doi.org/10.1016/j.ijft.2024.100859.
31. C. Shekhar, A. Gupta, N. Kumar, A. Kumar, and S. Varshney, "Transient Solution of Multiple Vacation Queue with Discouragement and Feedback," *Scientia Iranica*, vol. 29, no. 5, pp. 2567–2577, 2022.
32. T. T. Zin, A. S. T. Moe, C. N. Phyo, and P. Tin, "Fusion of Strategic Queueing Theory and AI for Smart City Telecommunication System," in *2024 IEEE 21st International Conference on Mobile Ad-Hoc and Smart Systems (MASS)*, pp. 653–657, 2024, https://doi.org/10.1109/MASS62177.2024.00104.
33. X. Li, "Performance Evaluation of M/M/1 Queueing Models in Cloud Computing Environments," *Highlights in Science, Engineering and Technology*, vol. 85, pp. 841–848, 2024, https://doi.org/10.54097/1xcw8y74.
34. B. Saini, D. Singh, and D. Sharma, "Exploring the Role of Queueing Theory in Manufacturing: An Analytical Study," *International Research Journal on Advanced Engineering and Management (IRJAEM)*, vol. 2, no. 3, pp. 256–266, 2024, https://doi.org/10.47392/irjaem.2024.0039.
35. A. Anagha, and S. Sreelatha, "Relevance of Queueing Theory in a Supermarket: A Case Study," in *2023 International Conference on Circuit Power and Computing Technologies (ICCPCT)*, pp. 550–553, 2023, https://doi.org/10.1109/ICCPCT58313.2023.10245273.
36. N. Mathavavisakan, and K. Indhira, "Nonlinear Metaheuristic Cost Optimization and ANFIS Computing of Feedback Retrial Queue with Two Dependent Phases of Service Under Bernoulli Working Vacation," *Int J Mod Phys B*, vol. 38, no. 30, p. 2440004, https://doi.org/10.1142/s0217979224400046.
37. K. Samouylov, O. Dudina, and A. Dudin, "Analysis of Multi-server Queueing System with Flexible Priorities," *Mathematics*, vol. 11, no. 4, p. 1040, 2023, https://doi.org/10.3390/math11041040.
38. R. Roy, A. Biswas, and A. Pal, "Queues with Resetting: A Perspective," *Journal of Physics: Complexity*, vol. 5, p., 2024, https://doi.org/10.1088/2632-072X/ad3e5a.
39. S. Singh, S. Acharya, F. Cruz, and R. Quinino, "Bayesian Inference and Prediction in a Queueing System," *Commun Stat Theory Methods*, vol. 52, pp. 8844–8864, 2023, https://doi.org/10.1080/03610926.2022.2076120.

40. R. Hasan, O. I. Obaid, A. I. M. AlQassab, H. Jasim, and S. A. Dheyab, "Mapping Web Service Characteristics to Queueing Theory Models for Performance Analysis," *Babylonian Journal of Networking*, pp. 98–110, 2024, https://doi.org/10.58496/bjn/2024/011.
41. Y. Wang, "Research on the Queuing Theory Based on M/M/1 Queuing Model," *Highlights in Science, Engineering and Technology*, vol. 61, pp. 80–87, 2023, https://doi.org/10.54097/hset.v61i.10276.
42. B. Kushvaha, D. Das, and A. Tamuli, "Single Server Queueing Model with Poisson Arrival and Gamma Service Time Distribution," *Global Journal of Computer Sciences: Theory and Research*, 2024, https://doi.org/10.18844/gjcs.v14i1.9324.

Chapter 2

Parametric Investigation of Retrial Queueing Systems with Customer Feedback Mechanism and Independent Abandonments

Shreekant Varshney, Manthan Shah, Bhasuru Abhinaya Srinivas, Ankit Deshmukh, and Anurag Kumar Srivastava

2.1 INTRODUCTION

The queueing modeling presents a rigorous mathematical foundation for understanding the dynamics of waiting queue phenomena in service-oriented systems. By systematically modeling the customer arrivals, service operations, and overall system capacity, the queueing mechanism improves the estimation of performance measures such as wait times, queue lengths, and service efficiency, hence boosting decision-making in operational management and the utilization of resources. This theoretical approach is essential for identifying constraints and enhancing customer experiences across numerous sectors. Modern service environments periodically encounter scenarios where blocked customers demand service retrials at random periods, illustrating real-world operational dynamics. The increasing complexity of customer behavioral patterns in modern-day service industries requires the development of advanced analytical frameworks. This investigation analyzes a queueing system including retrials, feedback mechanisms, and individual impatience, addressing fundamental challenges in service optimization and resource allocation. While retrial queueing systems were established primarily in telecommunication applications, examining interrupted call behaviors, their importance has increased substantially across numerous sectors. Nowadays, modern applications utilizing such queueing patterns include computer networks, healthcare systems, electric vehicle (EV) charging infrastructure, and customer service operations. The relationships between retrial phenomena and feedback mechanisms, where customers may require multiple service attempts, provide complex system dynamics requiring prudent mathematical investigation. Customer impatience shown by abandoning behavior causes novel analytical barriers, notably in performance measurement and system optimization.

Service systems operating under queueing assumptions demonstrate extensive correlations among arrival patterns, service mechanisms, and customer behaviors. Service providers must efficiently manage numerous operational

DOI: 10.1201/9781003642091-2

aspects: facilitating customer retrials, addressing feedback demands, and reducing abandonment impacts. Understanding these interrelationships becomes critical for developing operational strategies that improve service quality and resource utilization. The present research proposes an analytical framework incorporating various behavioral features while preserving practical applicability. Through rigorous mathematical analysis along with comprehensive performance assessment, the proposed investigation integrates theoretical foundations with real-world applications within modern service environments. A schematic diagram demonstrating the working procedure for the queueing infrastructure based on the suggested queueing framework is illustrated in Figure 2.1.

The evolution of queueing-based stochastic modeling has developed significantly, characterized by both analytical developments and practical applications. Pioneering efforts in queueing modeling were developed by Gupta [1], establishing an environment for subsequent applications of customer behavior. For instance, Li et al. [2] investigated queue-switching behaviors, underlining the necessity of identifying customer dynamics. Analogously, Strzęciwilk [3] developed timed Petri nets to improve modeling precision, while Bekker et al. [4] examined integrated routing methods, contributing to the adaptability of queueing frameworks. Moreover, the combination of artificial intelligence (AI) with the theory of queues has offered novel opportunities for research. This relationship is well-documented by Zin et al. [5] and Saini et al. [6], who demonstrated applications throughout smart city structures and manufacturing systems. Furthermore, healthcare applications also gained substantial attention, with contributions provided by Liu et al. [7], Chocron et al. [8], Xie et al. [9], and Wang et al. [10] emphasizing patient flow optimization and the management of large-scale vaccination programs.

Recent research has demonstrated substantial breakthroughs in addressing the complexity of service systems, specifically referring to feedback, unreliable servers, and individual impatience. Developing a Markovian

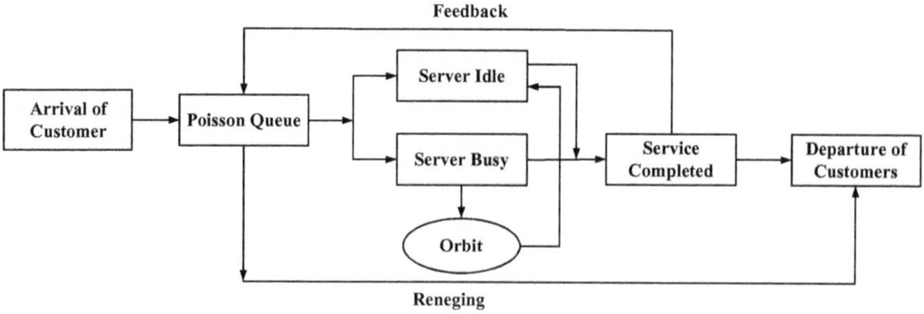

Figure 2.1 Flow pattern representing the service provisioning in the queueing network based on the proposed queueing framework.

framework for EV charging networks, Shah et al. [11] successfully integrated considerations of abandonment and feedback procedure. Varshney et al. [12] and Nayak et al. [13] focused their research on optimizing performance within systems incorporating unreliable servers by applying nature-inspired algorithms. Moreover, Shekhar et al. [14–16] conducted detailed assessments of multi-server queueing systems, taking into consideration factors such as vacation policies, reneging, and feedback mechanisms.

Additionally, Varshney et al. [17] enhanced these models to cover renewable energy and EV infrastructure applications, highlighting the adaptability of the queueing procedure. The appropriate transient solution methodologies and optimal control approaches via meta-heuristic techniques were examined by Shekhar et al. [18]. The importance of unified stochastic models and advanced optimization techniques in improving service quality and economic efficiency within modern service environments was highlighted by a number of studies, including those by Varshney et al. [19] and Shekhar et al. [20, 21]. These studies further emphasized the integration of matrix-analytic methods with cost-optimization strategies.

Modern investigation progressively emphasizes particular applications based on economic factors. The EV charging infrastructure has emerged as an important domain of attention, with substantial contributions from Varshney et al. [17, 22] addressing working vacations and arrival control procedures. In the domain of web service performance modeling, significant advancements were made by Hasan et al. [23] and Wang [24]. Additionally, studies conducted by Kushvaha et al. [25] and Li et al. [26] examined service distributions and hybrid priority systems, thereby enhancing the queueing perspectives with numerous concepts and approaches. As summarized in Tables 2.1 and 2.2, the collective findings of numerous authors illustrate the critical interrelationship between multiple solution methodologies and optimal operational strategies, indicating a paradigm shift toward more efficient queueing models that can better address the complexities of modern service environments.

The combination of diverse behavioral characteristics has created advanced analytical frameworks. Multi-server systems with complex characteristics have been explored extensively (cf. Samouylov et al. [28], Varshney et al. [19]). Renewable energy applications gained advantages from improved queueing models through Varshney et al. [29], combining feedback mechanisms and hyper-exponential service provisioning. Moreover, transient solutions have been derived by numerous researchers (cf. Shekhar et al. [30], Singh et al. [31], Roy et al. [32]), examining different aspects of customer discouragement and system dynamics.

Recent advances have focused on optimization and practical implementation methodologies. The supermarket queueing procedures have been investigated by Anagha and Sreelatha [33], while applications for cloud computing have been explored by Li [34]. The integration of partial server breakdowns and recovery procedures has been examined thoroughly by

Table 2.1 Recent Studies Based on Queueing Modeling with Multiple Attributes and Optimization Approaches.

Authors	*Methodology*	*Key Findings*
Varshney et al. [17]	Queueing-based stochastic modeling, evolutionary algorithms	Optimal converging outcomes through comparative investigation of multiple evolutionary algorithms for hybrid electric vehicle charging infrastructure
Mathavavisakan and Indhira [27]	Nonlinear metaheuristic optimization, adaptive neuro-fuzzy inference system computing	Cost optimization is performed by employing particle swarm optimization and artificial bee colony techniques
Shekhar et al. [18]	Matrix-geometric approach, meta-heuristic optimization	Cost optimization is accomplished by integrating the retention policy of reneged customers in finite capacity systems
Samouylov et al. [28]	Mathematical analysis, numerical illustration	Effective management of associated request flows with priority-based extraction rates
Shah et al. [11]	Markovian modeling, queueing networks	Effective solutions for EV charging infrastructure considering queueing management and abandonments

Table 2.2 Applications of the Queueing Hypothesis in Various Domains with Distinctive Characteristics.

Authors	*Application Domain*	*Major Contributions*
Zin et al. [5]	Smart city and telecommunication systems	Novel approach combining queueing procedures with AI methodologies for system optimization
Saini et al. [6]	Manufacturing systems	Enhanced efficiency through analysis of arrival rates and system capacities
Chocron et al. [8]	Healthcare sector	Advanced methodologies for managing patient flow and resource allocation
Li et al. [26]	Priority-based service systems	More extensive achievability area that allows more possibilities of waiting time configurations
Anagha & Sreelatha [33]	Supermarket and efficient operations	Practical insights for staffing and resource allocation during peak hours

Varshney et al. [35], integrating service pressure conditions and threshold-based control strategies. Meta-heuristic approaches for system optimization have gained popularity (cf. Varshney et al. [36], Shah et al. [11], Mathavavisakan and Indhira [27]), establishing new benchmarks for performance enhancement in complex queueing networks focusing on economic considerations and sensitivity assessments.

Furthermore, based on an extensive review of the existing literature, several significant gaps have been observed:

- Insufficient analysis of combined retrial–feedback mechanisms with abandoning behavior of customers.
- Limited exploration of steady-state solutions for complex queueing systems incorporating multiple behavioral aspects.
- Lack of comprehensive performance optimization frameworks considering numerous critical system parameters.
- Inadequate investigation of cost-effective operational strategies in the presence of retrials and customer impatience.

Similarly, the primary objectives and specific novel contributions of the present investigation can be characterized as follows:

- To develop a comprehensive mathematical framework for finite retrial queues, incorporating both feedback mechanisms and customer abandonments.
- To establish closed-form expressions for steady-state probability distribution and critical system performance measures.
- To analyze the impact of retrial rates, feedback probability, and reneging parameters on system performance.
- To formulate and address the optimization problem for determining optimal system parameters.
- To provide practical insights for system designers and operators through extensive numerical investigations.

The overall structure of the current study is delineated as follows: Section 2.1 outlines the evolution of stochastic modeling through various assumptions and annotations. The governing set of Chapman–Kolmogorov differential equations is established utilizing the principles of birth and death processes and Markov chains that are described in Section 2.2. Section 2.3 highlights a straightforward methodological framework for determining the state probability distribution of the constructed stochastic model through the application of matrix solution techniques. Section 2.4 presents closed-form formulations for several system performance indicators to assess the overall effectiveness of queueing infrastructures. A thorough parametric analysis is performed utilizing various significant illustrations that authenticate the proposed modeling and study outcomes, as depicted by several graphs in

Section 2.5. Section 2.6 emphasizes numerous essential managerial recommendations that could aid system designers, decision-makers, researchers, and scholars in the process of making decisions. Lastly, Section 2.7 presents concluding perceptions, several potential future viewpoints, and an overview of possible extensions of the proposed queueing-based stochastic modeling.

2.2 MODEL ASSUMPTIONS AND NOTATIONS

This study considers a finite capacity retrial queueing system incorporating both feedback and customer reneging mechanisms. The system is assumed to have a maximum capacity K, with customers arriving according to an independent Poisson process with mean arrival rate λ, and they join the queue on a first-come-first-served basis. Service times are modeled following an exponential distribution with mean rate μ, and when the server is occupied, incoming customers must have to join the retrial orbit. From the retrial orbit, customers may retry for re-service after waiting for an exponentially distributed period with rate parameter γ. Upon service completion, a customer may require additional service due to unsatisfactory outcomes, leading to a reattempt into the system with probability p, or otherwise leave permanently with probability $q = (1 - p)$. Each customer in the orbit also has a specific patience level, and if this patience level (time duration) expires, the customer abandons the system; this abandoning behavior is modeled by an exponential distribution with rate parameter η. These assumptions and associated notations establish the analytical framework for evaluating system performance and understanding the correlations among retrial attempts, feedback procedures, and customer impatience.

Now, let us suppose that $J(t)$ is the state of the service provider at time instant t and $N(t)$ is the number of customers in the queueing network at time instant t. Thus, the process $\{(J(t), N(t)), t \geq 0\}$ comprises an irreducible continuous-time Markov chain (CTMC) with state-space $\Omega = \{(j,n), j = 0,1, \text{and } n = 0,1,\ldots,K\}$. Furthermore, integrating the input and output streams from system states and utilizing the notion of CTMC, the state transition diagram (cf. Figure 2.2) of the constructed model can be depicted as follows.

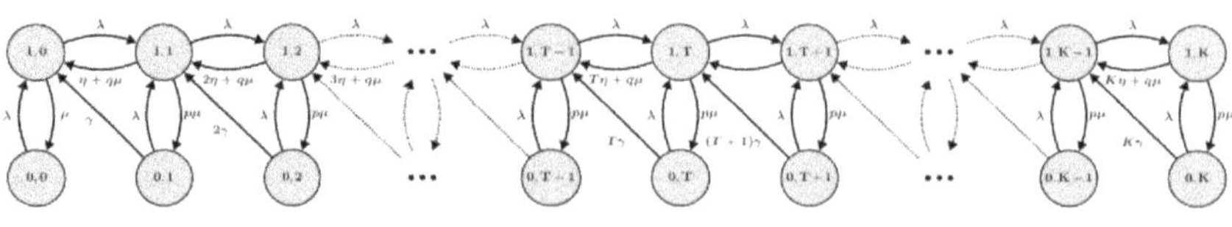

Figure 2.2 Transition state diagram associated with the proposed queueing network.

2.2.1 Differential-Difference Equations

In queueing modeling, differential-difference equations are essential because they establish a framework for examining dynamic systems in which continuous processes and specific occurrences interconnect with one another. These equations enable researchers to measure the unique characteristics of customer arrivals, service completions, and system conditions across varied time intervals, making them fundamental for understanding the behavior of complicated queueing networks. By combining both differential and difference factors, these equations facilitate the modeling of transient states and time-dependent phenomena, hence enhancing the accuracy of performance anticipates such as waiting times, queue lengths, and overall throughput of the service systems. Consequently, their implementation is essential for improving service efficiency and optimizing decision-making in diverse operational situations. Furthermore, the differential-difference equations associated with the developed queueing framework with retrial policy, independent abandonments, and the feedback mechanism can be characterized as follows:

$$\frac{dP_{0,0}(t)}{dt} = -\lambda P_{0,0}(t) + \mu P_{1,0}(t), \tag{2.1}$$

$$\frac{dP_{0,n}(t)}{dt} = -(\lambda + n\gamma)P_{0,n}(t) + p\mu P_{1,n}(t), \quad 1 \le n \le K-1, \tag{2.2}$$

$$\frac{dP_{0,K}(t)}{dt} = -(\lambda + K\gamma)P_{0,K}(t) + p\mu P_{1,K}(t), \tag{2.3}$$

$$\frac{dP_{1,0}(t)}{dt} = -(\lambda + \mu)P_{1,0}(t) + (\eta + q\mu)P_{1,1}(t) + \lambda P_{0,0}(t) + \gamma P_{0,1}(t), \tag{2.4}$$

$$\begin{aligned} \frac{dP_{1,n}(t)}{dt} = {} & -(\lambda + \mu + n\eta)P_{1,n}(t) + ((n+1)\eta + q\mu)P_{1,n+1}(t) + \\ & \lambda P_{1,n-1}(t) + \lambda P_{0,n}(t), \quad 1 \le n \le K-1, \end{aligned} \tag{2.5}$$

$$\frac{dP_{1,K}(t)}{dt} = -(\mu + K\eta)P_{1,K}(t) + \lambda P_{1,K-1}(t) + \lambda P_{0,K}(t). \tag{2.6}$$

2.2.2 Steady-State Equations

In the equilibrium, i.e., as $t \to \infty$, the abovementioned system of equations (2.1)–(2.6) can be transformed to

$$\lambda P_{0,0} = \mu P_{1,0}, \tag{2.7}$$

$$(\lambda + n\gamma)P_{0,n} = p\mu P_{1,n}, \quad 1 \le n \le K-1, \tag{2.8}$$

$$(\lambda + K\gamma)P_{0,K} = p\mu P_{1,K}, \tag{2.9}$$

$$(\lambda + \mu)P_{1,0} = (\eta + q\mu)P_{1,1} + \lambda P_{0,0} + \gamma P_{0,1}, \tag{2.10}$$

$$(\lambda + \mu + n\eta)P_{1,n} = ((n+1)\eta + q\mu)P_{1,n+1} + \lambda P_{1,n-1} + \lambda P_{0,n}, \quad 1 \le n \le K-1, \tag{2.11}$$

$$(\mu + K\eta)P_{1,K} = \lambda P_{1,K-1} + \lambda P_{0,K}. \tag{2.12}$$

Furthermore, in addition to these steady-state equations, the normalization condition can also be mathematically represented as

$$\sum_{j=0}^{1}\sum_{n=0}^{K} P_{j,n} = 1. \tag{2.13}$$

2.3 MATRIX SOLUTION METHODOLOGY

The matrix-analytical procedure is a powerful methodology for identifying and developing stochastic models. This technique is commonly applied in dealing with varied queueing situations that require appropriate numerical interpretations. By applying the matrix technique, the abovementioned set of differential equations (2.7)–(2.12) can be transformed into the matrix expression $\pi\mathbf{Q} = 0$, where π is the steady-state probability vector of dimension $(2K+2)$ and $\mathbf{Q}$ is the generator matrix corresponding to all the state probabilities. The structure of the correlated block tri-diagonal rate matrix $\mathbf{Q}$ correlated with the underlying Markov chain and of dimension $(2K+2)\times(2K+2)$ can be illustrated as

$$\mathbf{Q} = \begin{bmatrix}
\mathbf{A}_0 & \mathbf{B} & 0 & \cdots & 0 & 0 & 0 & \cdots & 0 & 0 & 0 \\
\mathbf{C}_1 & \mathbf{A}_1 & \mathbf{B} & \cdots & 0 & 0 & 0 & \cdots & 0 & 0 & 0 \\
0 & \mathbf{C}_2 & \mathbf{A}_2 & \cdots & 0 & 0 & 0 & \cdots & 0 & 0 & 0 \\
\vdots & \vdots & \vdots & \ddots & \vdots & \vdots & \vdots & \ddots & \vdots & \vdots & \vdots \\
0 & 0 & 0 & \cdots & \mathbf{A}_{T-1} & \mathbf{B} & 0 & \cdots & 0 & 0 & 0 \\
0 & 0 & 0 & \cdots & \mathbf{C}_T & \mathbf{A}_T & \mathbf{B} & \cdots & 0 & 0 & 0 \\
0 & 0 & 0 & \cdots & 0 & \mathbf{C}_{T+1} & \mathbf{A}_{T+1} & \cdots & 0 & 0 & 0 \\
\vdots & \vdots & \vdots & \ddots & \vdots & \vdots & \vdots & \ddots & \vdots & \vdots & \vdots \\
0 & 0 & 0 & \cdots & 0 & 0 & 0 & \cdots & \mathbf{A}_{K-2} & \mathbf{B} & 0 \\
0 & 0 & 0 & \cdots & 0 & 0 & 0 & \cdots & \mathbf{C}_{K-1} & \mathbf{A}_{K-1} & \mathbf{B} \\
0 & 0 & 0 & \cdots & 0 & 0 & 0 & \cdots & 0 & \mathbf{C}_K & \mathbf{A}_K
\end{bmatrix},$$

where

$$\mathbf{A}_0 = \begin{bmatrix} -\lambda & \lambda \\ \mu & -(\lambda+\mu) \end{bmatrix}, \quad \mathbf{B} = \begin{bmatrix} 0 & 0 \\ 0 & \lambda \end{bmatrix},$$

$$\mathbf{A}_n = \begin{bmatrix} -(\lambda+n\eta) & \lambda \\ p\mu & -(\lambda+n\eta+\mu) \end{bmatrix}, \quad 1 \le n \le K-1,$$

$$\mathbf{A}_K = \begin{bmatrix} -(\lambda+K\gamma) & \lambda \\ p\mu & -(K\eta+\mu) \end{bmatrix} \text{and } \mathbf{C}_n = \begin{bmatrix} 0 & n\gamma \\ 0 & (n\eta+q\mu) \end{bmatrix},$$
$$1 \le n \le K-1.$$

Here, the state probability vector associated with the rate matrix Q in the equilibrium is represented by π, which can be partitioned into $\pi = [\pi_0, \pi_1, \pi_2, \cdots, \pi_{T-1}, \pi_T, \pi_{T+1}, \cdots, \pi_{K-1}, \pi_K]$, where sub-vectors $\pi_n = [P_{0,n}, P_{1,n}]$ for $n = 0,1,2,\ldots,K$ are of dimension (1×2). Let us suppose that vector e represents a column vector of dimension $(2\ K+2)$, all of which elements as ones. Now, utilizing and solving the homogeneous system of linear equations $\pi\mathrm{Q} = 0$ and employing the normalizing condition $\pi\mathrm{e} = 1$, the steady-state probabilities associated with the abovementioned matrix equations can be easily depicted numerically with the help of MATLAB (2024b) software.

2.4 SYSTEM PERFORMANCE MEASURES

In queueing modeling, the demand for varied system performance measurements is crucial in establishing efficient management strategies and boosting service efficiency. Important metrics such as the expected number of customers in the orbit, average rate of abandonment of each individual, overall throughput of the system, and server utilization offer significant insights into system operation and operational performance. By assessing these indicators, researchers and practitioners can identify challenges, optimize resource allocation, and enhance service processes to meet fluctuating demand levels. Furthermore, combining several performance indicators allows for a more thorough understanding of alternatives throughout the system, enabling accurate decision-making that ultimately leads to improved customer experiences and operational effectiveness in various service circumstances.

1) Expected number of customers in the orbit:

$$L_S = \sum_{j=0}^{1}\sum_{n=0}^{K} nP_{J,n}. \tag{2.14}$$

2) Expected number of individuals attempting for their turn from the orbit:

$$L_O = \sum_{n=1}^{K} nP_{0,n}. \tag{2.15}$$

3) Average rate of abandonment of each individual:

$$ARR = \sum_{n=1}^{K} (n\eta)P_{1,n}. \tag{2.16}$$

4) Average rate of individuals rejoining the system to complete the unsatisfactory service:

$$AFM = \sum_{n=1}^{K} (p\mu)P_{1,n}. \tag{2.17}$$

5) Average retrial rate of individuals in the system:

$$AR = \sum_{n=1}^{K} (n\gamma)\ P_{0,n}. \tag{2.18}$$

6) Overall throughput of the system:

$$\tau_p = \sum_{n=0}^{K} \mu P_{1,n}. \tag{2.19}$$

7) Expected waiting time of customers in the orbit:

$$W_S = \left(\frac{L_S}{\lambda}\right). \tag{2.20}$$

2.5 NUMERICAL RESULTS

In this section, several numerical illustrations are provided to validate the proposed modeling based on the queueing theoretic framework. For illustration purposes, we selected the system parameters as $K = 15, \lambda = 2.5, \mu = 3.0, \eta = 0.1, \gamma = 2.0, p = 0.1$, and $q = 0.9$. The numerical results derived from the suggested modeling are shown through several line plots, exhibiting the system's behavior under varied operational configurations. These parameters have been carefully selected to simulate realistic conditions in service systems demonstrating retrial, feedback, and reneging behaviors, considering typical customer behavior, demand patterns, and operational constraints.

The interrelationships between the arrival rate (λ) and the expected number of customers in the orbit $E(O)$ exhibiting diverse patterns across varying system parameter values are depicted in Figure 2.3. The impact of

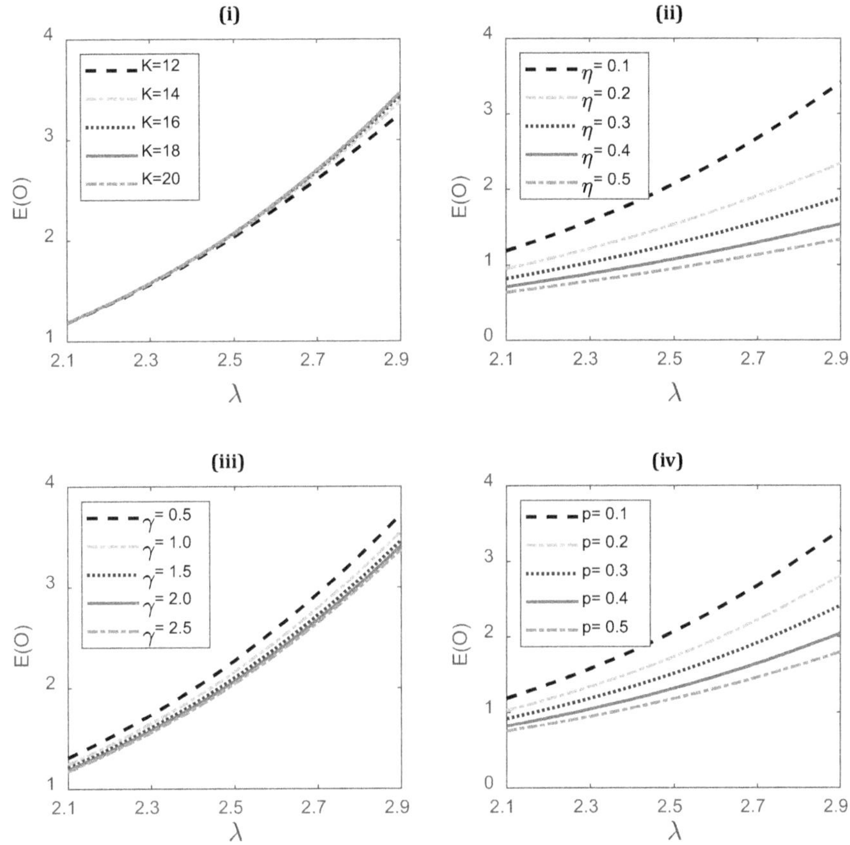

Figure 2.3 Impact of arrival rate (λ) and other system parameters on the expected number of customers in the orbit.

system capacity (K) is demonstrated in Figure 2.3(i), where different capacity values ($K = 12, 14, 16, 18, 20$) exhibit nearly identical linear growth patterns, with $E(O)$ increasing from approximately 1.0 to 3.5 as λ increases from 2.1 to 2.9, indicating that system capacity has minimal impact on expected orbit population under stable operating conditions. In contrast, Figure 2.3(ii) demonstrates that the reneging rate (η) significantly impacts orbit population, with lower reneging values ($\eta = 0.1$) resulting in substantially higher $E(O)$ values (reaching approximately 3.3 at $\lambda = 2.9$), while higher reneging rates ($\eta = 0.5$) maintain lower orbit populations (approximately 1.3 at $\lambda = 2.9$), confirming that customer impatience serves as an effective natural control mechanism for orbit congestion. Figure 2.3(iii) reveals that decreased retrial rates (γ) lead to higher orbit populations, with the lowest rate ($\gamma = 0.5$) resulting in $E(O)$ reaching approximately 3.7 at $\lambda = 2.9$, while the highest rate ($\gamma = 2.5$) maintains $E(O)$ at approximately 3.2 under the same arrival conditions, indicating that faster retrials actually reduce orbit congestion by expediting service attempts. Figure 2.3(iv) shows that higher feedback probabilities ($p = 0.1$) generate substantially larger orbit populations ($E(O) \approx 3.5$ at $\lambda = 2.9$) compared to lower probabilities ($p = 0.5$, $E(O) \approx 2.0$ at $\lambda = 2.9$), demonstrating that service quality significantly influences orbit management.

Figure 2.4 illustrates how the correlation between service rate (μ) and predicted orbit population $E(O)$ changes with different system characteristics. Figure 2.4(i) demonstrates that system capacity (K) configurations follow nearly identical declining trajectories as μ increases from 2.1 to 2.9, with $E(O)$ decreasing from approximately 5.0 to 2.0, further confirming the minimal influence of structural capacity on operational performance. Customer impatience significantly modulates orbit congestion as illustrated in Figure 2.4(ii), where lower reneging rates ($\eta = 0.1$) maintain consistently higher $E(O)$ values (decreasing from 5.5 to 2.5) compared to higher

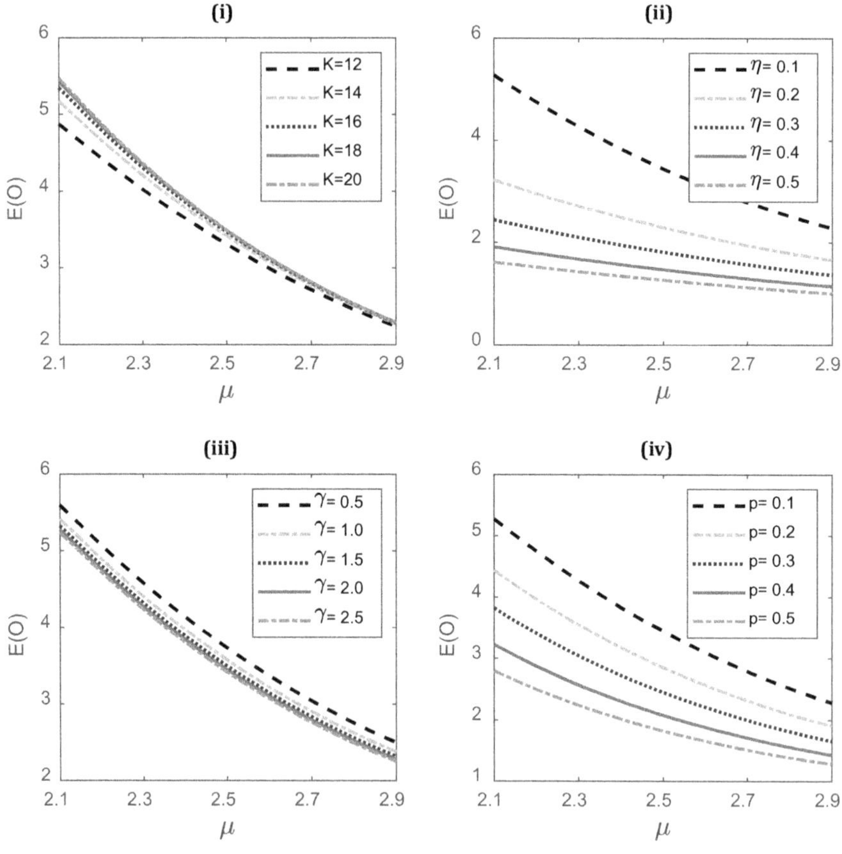

Figure 2.4 Impact of service rate (μ) and other system parameters on the expected number of customers in the orbit.

reneging rates ($\eta = 0.5$, decreasing from 1.5 to 1.0) across all service rate values. Figure 2.4(iii) reveals that retrial rate (γ) has a decreasing impact as service efficiency improves, with lower rates ($\gamma = 0.5$) showing higher initial $E(O)$ values (5.5 at $\mu = 2.1$) but converging with higher rates as μ approaches 2.9. The profound influence of service quality is demonstrated in Figure 2.4(iv), where lower feedback probabilities ($p = 0.5$) consistently maintain smaller orbit populations across all service rates, with the separation between curves remaining relatively constant throughout the μ range. These patterns align with practical service system dynamics, where operational efficiency typically exerts greater influence than structural capacity, and customer behavioral responses, particularly service satisfaction and patience thresholds, critically determine system congestion.

The feedback probability (q) demonstrates substantial influence on the expected orbit population $E(O)$ across various parameter configurations as depicted in Figure 2.5. Examining Figure 2.5(i), system capacity (K) exhibits

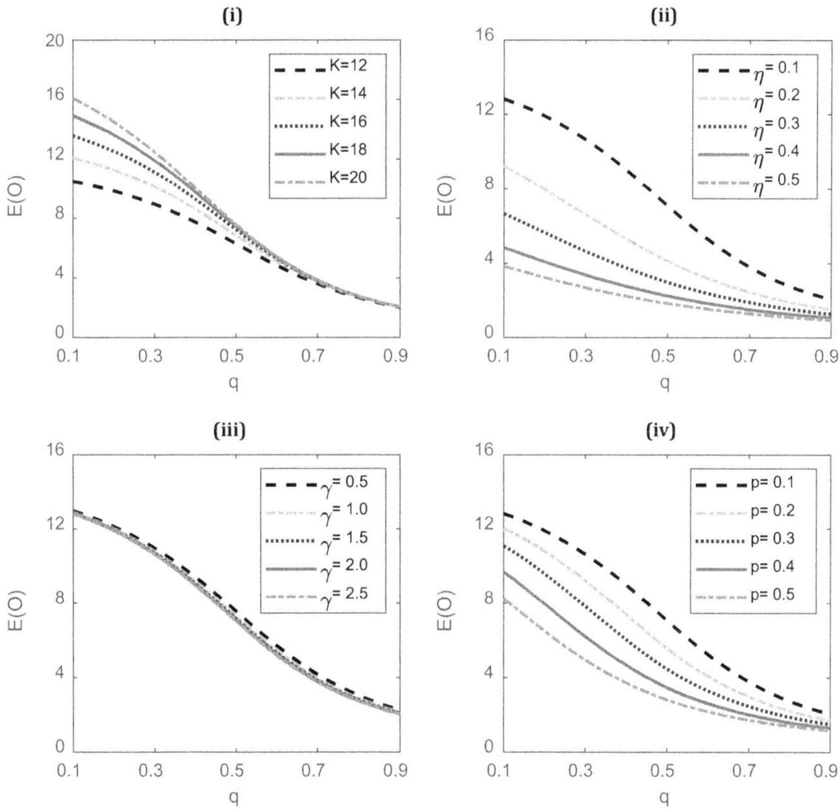

Figure 2.5 Impact of feedback probability (q) and other system parameters on the expected number of customers in the orbit.

divergent impacts at lower q values with higher capacities ($K = 20$) reaching $E(O) \approx 16$, although these differences diminish as q increases, with all configurations converging near $E(O) \approx 2$ at $q = 0.9$. In addition, Figure 2.5(ii) illustrates how the reneging rate (η) modifies the impact of feedback probability, with lower values ($\eta = 0.1$) maintaining consistently higher orbit populations across the q range. Interestingly, Figure 2.5(iii) reveals that retrial rate (γ) has a minimal differential impact on how feedback probability affects orbit population, with all configurations following nearly identical trajectories. Figure 2.5(iv) shows that lower service quality (higher p values) maintains higher orbit populations across most q values, although all configurations converge as q approaches 0.9. These patterns reflect real-world service dynamics, where customer departure decisions effectively counter structural inefficiencies.

Figure 2.6 illustrates the impact of arrival rate (λ) on system throughput (τ_p) across varying system parameters. Figure 2.6(i) demonstrates that

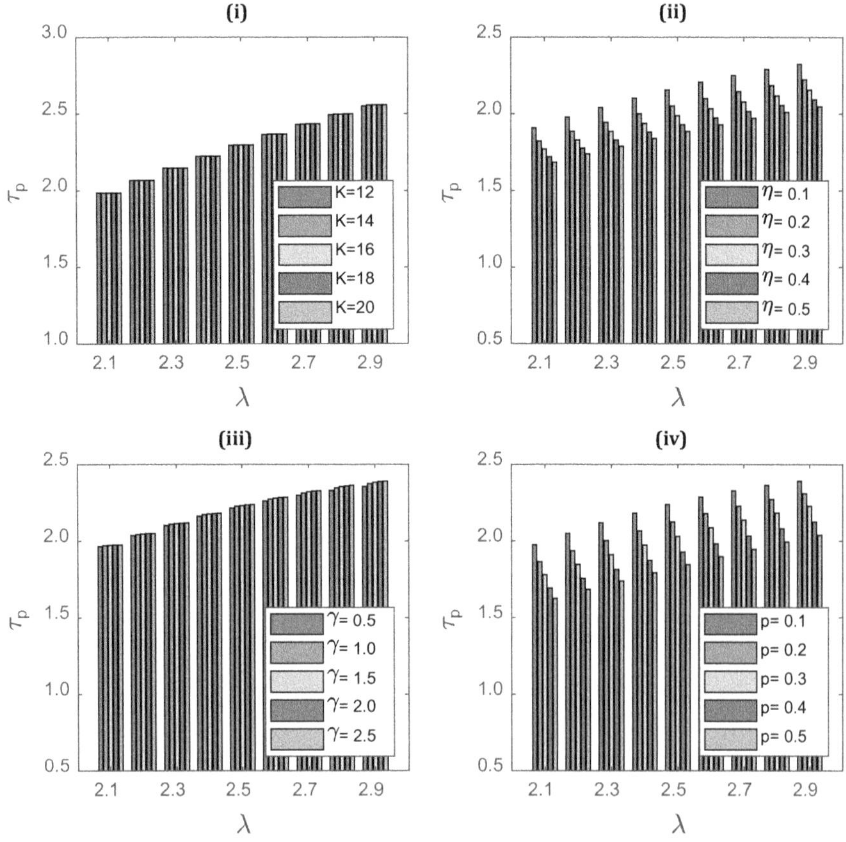

Figure 2.6 Variation in the overall throughput of the system with regard to arrival rate (λ) and other system parameters.

system capacity (K) has minimal influence on throughput, with all capacity configurations exhibiting nearly identical increasing trajectories as λ increases from 2.1 to 2.9, where τ_p increases from approximately 2.0 to 2.6. Customer impatience significantly affects system throughput as shown in Figure 2.6(ii), where higher reneging rates $(\eta = 0.5)$ consistently yield lower throughput values (ranging from 1.7 to 2.1) compared to lower reneging rates ($\eta = 0.1$, ranging from 1.9 to 2.3) across all arrival rates. Figure 2.6(iii) reveals that the retrial rate (γ) has a negligible impact on throughput, with all values producing virtually identical performance across the entire λ spectrum. The service quality parameter depicted in Figure 2.6(iv) shows that lower feedback probabilities $(p = 0.1)$ maintain consistently higher throughput compared to higher probabilities $(p = 0.5)$ across all arrival rates, reinforcing the critical influence of service quality on overall system efficiency.

The relationship between service rate (μ) and system throughput (τ_p) for various system characteristics is displayed in Figure 2.7. In Figure 2.7(i), it is revealed that system capacity (K) has a negligible impact on system throughput, with all capacity configurations showing virtually identical increasing trajectories as μ increases from 2.1 to 2.9, where τ_p increases from approximately 1.9 to 2.3. The effect of customer impatience is demonstrated in Figure 2.7(ii), where lower reneging rates $(\eta = 0.1)$ consistently produce higher throughput values (ranging from 1.4 to 1.45) compared to higher reneging rates ($\eta = 0.5$, ranging from 1.25 to 1.35) across all service rates. Figure 2.7(iii) indicates that the retrial rate (γ) has minimal influence on throughput performance, with all values yielding nearly identical results as the service rate increases. The significant impact of service quality is illustrated in Figure 2.7(iv), where lower feedback probabilities $(p = 0.1)$ maintain substantially higher throughput (approximately 1.45) compared to higher feedback probabilities ($p = 0.5$, approximately 1.2) throughout the entire μ spectrum, highlighting how service quality fundamentally determines overall system efficiency regardless of service rate improvements.

Similarly, Figure 2.8 illustrates the significant impact of feedback probability (q) on system throughput (τ_p) across various system parameters. Figure 2.8(i) demonstrates that system capacity (K) has minimal influence on throughput, with all capacity configurations exhibiting nearly identical declining trajectories as q increases from 0.1 to 0.9, where τ_p decreases substantially from approximately 2.9 to 2.1. Customer impatience markedly affects system throughput as shown in Figure 2.8(ii), where lower reneging rates $(\eta = 0.1)$ consistently yield higher throughput values (decreasing from 2.4 to 1.4) compared to higher reneging rates ($\eta = 0.5$, decreasing from 2.2 to 1.2) across all feedback probabilities. Figure 2.8(iii) reveals that the retrial rate (γ) has a moderate impact on throughput at lower q values but becomes negligible as q increases, with all curves converging toward 1.5 at $q = 0.9$. The service quality parameter depicted in Figure 2.8(iv) shows that lower feedback probabilities $(p = 0.1)$ maintain consistently higher throughput compared to higher feedback probabilities $(p = 0.5)$ across all q values,

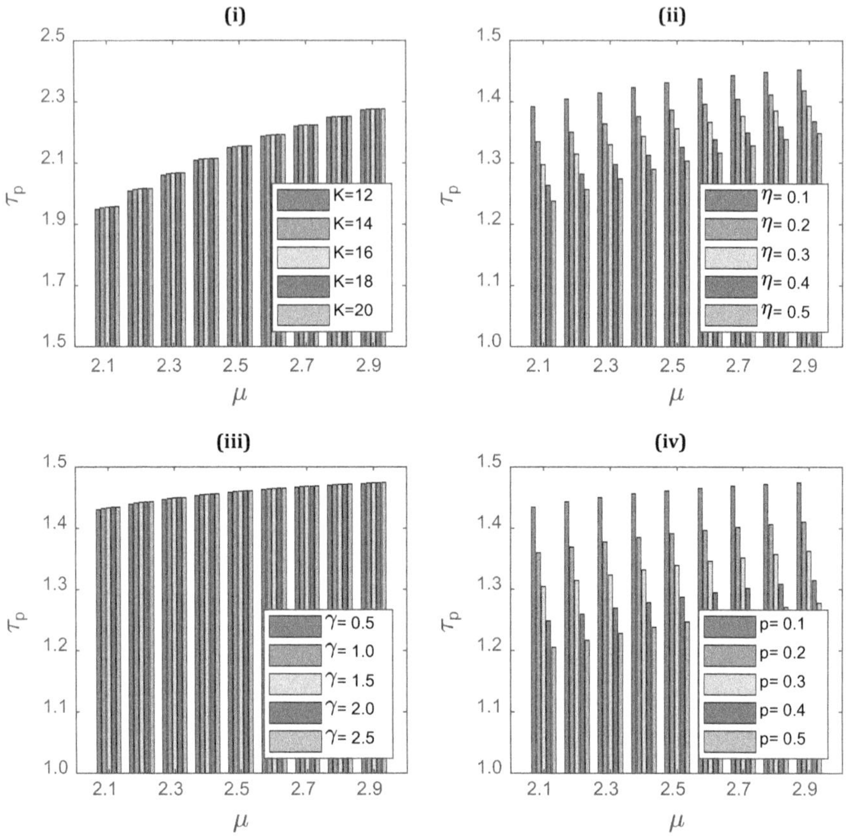

Figure 2.7 Variation in the overall throughput of the system with regard to service rate (μ) and other system parameters.

with the performance gap widening as q increases, highlighting the compounding negative effect of both feedback mechanisms on system efficiency. The observed relationships align with practical service system dynamics, where increasing feedback probability represents customers requiring multiple service attempts, naturally reducing overall throughput regardless of system capacity. These patterns reflect real-world scenarios where service quality fundamentally determines system efficiency more than structural parameters, and higher customer abandonment rates (reneging) consistently decrease throughput across all configurations. The minimal impact of retrial intensity at higher feedback probabilities demonstrates how service quality issues ultimately surpass customer persistence in determining system performance, a phenomenon commonly observed in competitive service environments.

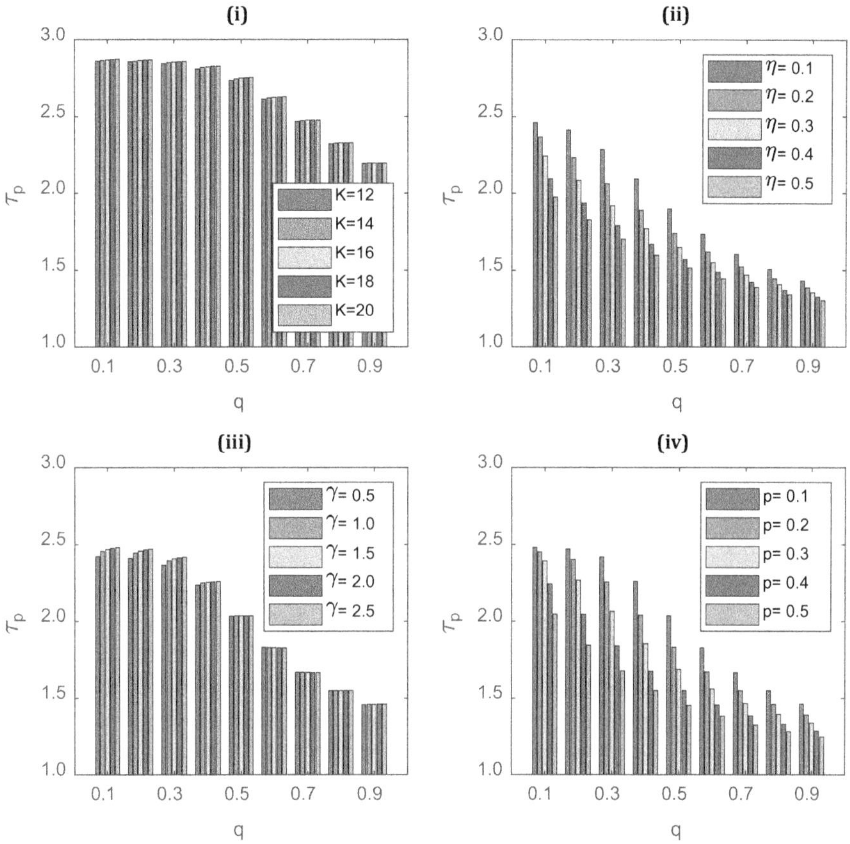

Figure 2.8 Variation in the overall throughput of the system with regard to feedback probability (q) and other system parameters.

2.6 SIGNIFICANT MANAGERIAL INSIGHTS

The analytical findings from the current study provide significant insights for optimal service system management in queueing-based stochastic modeling, focusing on customer abandonment, feedback mechanisms, and retrial policies. The investigation emphasizes the substantial impact of retrial behaviors on system congestion, particularly under high arrival rates. To reduce peak-time congestion, system managers should consider establishing organized, scheduled mechanisms that permit a more equitable distribution of retrials. Moreover, presenting customers with anticipated waiting times is essential for managing expectations and minimizing abandonment rates.

The inherent feedback mechanism in the service system provides service cycles that promote system utilization; nevertheless, it may also result in unpredictable service completion intervals. Consequently, decision-makers are recommended to build specialized processes for customers requiring numerous service initiatives that promote a first-time solution. Additionally, developing independent service channels for prospective customers and individuals in feedback cycles can further streamline service delivery. In addition, customer departure represents a substantial loss of commercial prospects and can seriously deteriorate the credibility of a business. Our sensitivity analysis suggests that minimizing reneging behaviors may significantly increase overall throughput. Practical interventions to address this include the development of queue-monitoring technologies that indicate service progress, deploying engagement activities during waiting periods, and giving alternate service options during peak congestion.

Moreover, the developed optimization model offers decision-making assistance to construct optimal service strategies that successfully balance operating costs with abandonment consequences. Furthermore, strategic investments in service enhancement, technological advances, and resource allocation between primary service and retrial management systems are proposed. The comparative analysis between critical operational parameters demonstrates fundamental relationships, particularly the impact of service rate alterations on waiting time reductions. Besides, implementing feedback minimization measures shows that it is a more effective strategy for boosting system throughput than making comparable investments in capacity development. Additionally, dynamically reallocating resources based on real-time monitoring of orbiting customers can considerably boost overall system performance. Collectively, these insights empower managers and system designers to develop robust service systems that accommodate complex customer patterns while improving resource utilization.

2.7 CONCLUSION AND FUTURE PERSPECTIVES

This study presents a comprehensive mathematical framework for analyzing queueing systems with retrials, feedback mechanisms, and customer reneging, three behavioral aspects that significantly impact contemporary service operations. Through rigorous analytical methods, we have derived the closed-form expressions for steady-state probabilities and several critical performance measures, providing a solid foundation for system analysis and optimization. The investigation revealed that the interaction between retrial phenomena and feedback cycles creates complex system dynamics that conventional queueing models fail to capture. Our integrated approach demonstrates how these mechanisms collectively influence waiting times, system length, and overall throughput of the system. Additionally, customer abandonment, while representing lost service opportunities, serves as a natural congestion control mechanism that prevents uncontrolled queueing growth under peak traffic conditions.

The research bridges theoretical queueing analysis with practical operational considerations, offering both analytical rigor and significant insights for service system design and management. The sensitivity analysis established quantitative relationships between system parameters and performance metrics, demonstrating that feedback probability has a pronounced impact on system congestion at high retrial rates, while reneging parameters significantly influence throughput and service completion probability. The optimization framework enables the identification of optimal system parameters that minimize total operational costs while maintaining acceptable service levels. By incorporating both economic considerations and service quality constraints, the model provides a practical decision-support tool for service system managers. The findings demonstrate that an integrated approach to managing these behavioral aspects yields superior performance compared to addressing each aspect independently.

Future research should focus on several potential directions: developing non-Markovian models with general service time distributions, investigating transient solutions for systems experiencing unexpected demand fluctuations, and incorporating strategic customer behavior models where decisions depend on observed system states. Computational approaches, including simulation frameworks and machine learning techniques, could advance real-time prediction of retrial patterns and adaptive resource allocation. Application domains could be expanded to healthcare systems with return appointments, renewable energy integration with intermittent availability, and EV charging infrastructure. Integrated optimization frameworks balancing multiple objectives and accommodating uncertainty in arrival patterns would enhance practical applicability, while examining sociotechnical considerations would provide valuable insights for system design. These extensions would strengthen both the theoretical foundations and practical applications of retrial queueing systems with feedback mechanisms and customer abandonments.

REFERENCES

1. S. M. Gupta, "Interrelationship Between Controlling Arrival and Service in Queueing Systems," *Computers & Operations Research*, vol. 22, no. 10, pp. 1005–1014, 1995.
2. Y. Li, Z. Cui, J. Ji, and J. Wei, "When Do People Switch Queues? An Empirical Study of Discretionary Queue Switching at a Physical Examination Center," *Decision Sciences*, vol. 56, no. 4, pp. 361–382, 2024, https://doi.org/10.1111/deci.12646.
3. D. Strzęciwilk, "Timed Petri Nets for Modeling and Performance Evaluation of a Priority Queueing System," *Energies (Basel)*, vol. 16, no. 23, p. 7690, 2023, https://doi.org/10.3390/en16237690.
4. R. Bekker, B. Bharti, L. Lan, and M. Mandjes, "A Queueing-Based Approach for Integrated Routing and Appointment Scheduling," *Eur. J. Oper. Res.*, vol. 318, pp. 534–548, 2023, https://doi.org/10.1016/j.ejor.2024.05.038.

5. T. T. Zin, A. S. T. Moe, C. N. Phyo, and P. Tin, "Fusion of Strategic Queueing Theory and AI for Smart City Telecommunication System," in *2024 IEEE 21st International Conference on Mobile Ad-Hoc and Smart Systems (MASS)*, pp. 653–657, 2024, https://doi.org/10.1109/MASS62177.2024.00104.
6. B. Saini, D. Singh, and D. Sharma, "Exploring the Role of Queueing Theory in Manufacturing: An Analytical Study," *International Research Journal on Advanced Engineering and Management (IRJAEM)*, vol. 2, no. 3, pp. 256–266, 2024, https://doi.org/10.47392/irjaem.2024.0039.
7. Y. Liu, L. Liu, T. Jiang, and X. Chai, "Strategic Queueing Behavior of Two Groups of Patients in a Healthcare System," *Mathematics*, vol. 12, no. 10, p. 1579, 2024, https://doi.org/10.3390/math12101579.
8. E. Chocron, I. Cohen, and P. Feigin, "Delay Prediction for Managing Multiclass Service Systems: An Investigation of Queueing Theory and Machine Learning Approaches," *IEEE Trans Eng Manag*, vol. 71, pp. 4469–4479, 2024, https://doi.org/10.1109/TEM.2022.3222094.
9. S. Xie *et al.*, "Enhancing Mass Vaccination Programs with Queueing Theory and Spatial Optimization," *Frontiers in Public Health*, vol. 12, p. 1440673, 2024.
10. X. Wang, H. Xie, and J. C. S. Lui, "Analyzing Queueing Problems via Bandits with Linear Reward and Nonlinear Workload Fairness," *IEEE Trans Mob Comput*, vol. 23, pp. 3410–3423, 2024, https://doi.org/10.1109/TMC.2023.3276158.
11. M. Shah, S. Varshney, B. A. Srinivas, and A. Bavisi, "Queueing Theoretical Approach and Markov Modeling Employed in Multiple Charging Stations of Electric Vehicle Networks," in *Proceedings of the 2024 International Conference on Sustainable Energy: Energy Transition and Net-Zero Climate Future, ICUE 2024*, 2024. https://doi.org/10.1109/ICUE63019.2024.10795574.
12. S. Varshney, C. Shekhar, V. Tiwari, and K. S. Pritam, "Different Variants of Unreliable Server: An Economic Approach," in *Modeling and Applications in Operations Research*, CRC Press, 2023, pp. 61–90. https://doi.org/10.1201/9781003462422-3.
13. J. Nayak, S. Varshney, and C. Shekhar, *Modeling and Applications in Operations Research*. CRC Press, Boca Raton, 2023.
14. C. Shekhar, S. Varshney, and A. Kumar, "Standbys Provisioning in Machine Repair Problem with Unreliable Service and Vacation Interruption," in *The Handbook of Reliability, Maintenance, and System Safety through Mathematical Modeling*. Elsevier, Amsterdam, 2021, pp. 101–133.
15. C. Shekhar, A. Kumar, and S. Varshney, "Parametric Nonlinear Programming for Fuzzified Queuing Systems with Catastrophe," *International Journal of Process Management and Benchmarking*, vol. 10, no. 1, pp. 69–98, 2020.
16. C. Shekhar, A. Kumar, S. Varshney, and S. I. Ammar, "Fault-Tolerant Redundant Repairable System with Different Failures and Delays," *Engineering Computations (Swansea, Wales)*, vol. 37, no. 3, pp. 1043–1071, 2020, https://doi.org/10.1108/EC-01-2019-0003.
17. S. Varshney, M. Shah, B. A. Srinivas, M. Gupta, and K. P. Panda, "Queueing Modeling and Optimization of Hybrid Electric Vehicle Infrastructures Using Evolutionary Algorithms," *International Journal of Thermofluids*, vol. 26, p. 101043, 2025, https://doi.org/10.1016/j.ijft.2024.101043.

18. C. Shekhar, S. Varshney, and A. Kumar, "Matrix-Geometric Solution of Multi-server Queueing Systems with Bernoulli Scheduled Modified Vacation and Retention of Reneged Customers: A Meta-Heuristic Approach," *Qual Technol Quant Manag*, vol. 18, no. 1, pp. 39–66, 2021.
19. S. Varshney, B. A. Srinivas, M. Gupta, M. Shah, and A. Bavisi, "Stochastic Modeling of Multiple-Server Charging Stations for Electric Vehicle Networks Using Feedback Strategies: A Queueing-Theoretic Approach," *International Journal of Thermofluids*, vol. 24, p. 100859, 2024, https://doi.org/10.1016/j.ijft.2024.100859.
20. C. Shekhar, A. Kumar, and S. Varshney, "Modified Bessel Series Solution of the Single Server Queueing Model with Feedback," *International Journal of Computing Science and Mathematics*, vol. 10, no. 3, pp. 313–326, 2019.
21. C. Shekhar, A. Kumar, and S. Varshney, "Load Sharing Redundant Repairable Systems with Switching and Reboot Delay," *Reliab Eng Syst Saf*, vol. 193, Jan. 2020, https://doi.org/10.1016/j.ress.2019.106656.
22. S. Varshney *et al.*, "Stochastic Modeling of Electric Vehicle Infrastructure Using Queueing-Theoretical Approach," *Results in Engineering*, vol. 25, p. 104149, Mar. 2025, https://doi.org/10.1016/j.rineng.2025.104149.
23. R. Hasan, O. I. Obaid, A. I. M. AlQassab, H. Jasim, and S. A. Dheyab, "Mapping Web Service Characteristics to Queueing Theory Models for Performance Analysis," *Babylonian Journal of Networking*, pp. 98–110, 2024, https://doi.org/10.58496/bjn/2024/011.
24. Y. Wang, "Research on the Queuing Theory Based on M/M/1 Queuing Model," *Highlights in Science, Engineering and Technology*, vol. 61, pp. 80–87, 2023, https://doi.org/10.54097/hset.v61i.10276.
25. B. Kushvaha, D. Das, and A. Tamuli, "Single Server Queueing Model with Poisson Arrival and Gamma Service Time Distribution," *Global Journal of Computer Sciences: Theory and Research*, 2024, https://doi.org/10.18844/gjcs.v14i1.9324.
26. Z. Li, M. Harchol-Balter, and A. Scheller-Wolf, "Hybrid Priority Queue and Its Applications," *ACM SIGMETRICS Performance Evaluation Review*, vol. 52, no. 2, pp. 12–14, 2024, https://doi.org/10.1145/3695411.3695417.
27. N. Mathavavisakan, and K. Indhira, "Nonlinear Metaheuristic Cost Optimization and ANFIS Computing of Feedback Retrial Queue with Two Dependent Phases of Service under Bernoulli Working Vacation," *Int J Mod Phys B*, vol. 38, no. 30, p. 2440004, 2023, https://doi.org/10.1142/s0217979224400046.
28. K. Samouylov, O. Dudina, and A. Dudin, "Analysis of Multi-Server Queueing System with Flexible Priorities," *Mathematics*, vol. 11, no. 4, p. 1040, 2023, https://doi.org/10.3390/math11041040.
29. S. Varshney *et al.*, "Optimal Management Strategies of Renewable Energy Systems with Hyperexponential Service Provisioning: An Economic Investigation," *Front Energy Res*, vol. 11, p. 1329899, 2023.
30. C. Shekhar, A. Gupta, N. Kumar, A. Kumar, and S. Varshney, "Transient Solution of Multiple Vacation Queue with Discouragement and Feedback," *Scientia Iranica*, vol. 29, no. 5, pp. 2567–2577, 2022.
31. S. Singh, S. Acharya, F. Cruz, and R. Quinino, "Bayesian Inference and Prediction in a Queueing System," *Commun Stat Theory Methods*, vol. 52, pp. 8844–8864, 2023, https://doi.org/10.1080/03610926.2022.2076120.

32. R. Roy, A. Biswas, and A. Pal, "Queues with Resetting: A Perspective," *Journal of Physics: Complexity*, vol. 5, no. 2, p. 021001, 2024, https://doi.org/10.1088/2632-072X/ad3e5a.
33. A. Anagha, and K. S. Sreelatha, "Relevance of Queueing Theory in a Supermarket: A Case Study," in *2023 International Conference on Circuit Power and Computing Technologies (ICCPCT)*, pp. 550–553, 2023, https://doi.org/10.1109/ICCPCT58313.2023.10245273.
34. X. Li, "Performance Evaluation of M/M/1 Queueing Models in Cloud Computing Environments," *Highlights in Science, Engineering and Technology*, vol. 85, pp. 841–848, 2024, https://doi.org/10.54097/1xcw8y74.
35. S. Varshney, S. Kaswan, M. Devanda, and C. Shekhar, "Finite Capacity Service System with Partial Server Breakdown and Recovery Policy: An Economic Perspective," *J Syst Sci Syst Eng*, pp. 1–31, 2024.
36. S. Varshney, M. M. Acharya, Z. J. Kanjibhai, M. Pushkarna, W. El-Shafai, and I. Zaitsev, "Economic and Sensitivity Analyses of Controllable Queues with Customer Feedback and Impatience Using Markovian Modeling," *Journal of Mathematics*, vol. 2025, no. 1, Jan. 2025, https://doi.org/10.1155/jom/9648114.

Chapter 3

Multi-Choice Multi Objective Fuzzy Stochastic Transportation Problem

P. K. Rout, S. Nanda, M. M. Acharya, and Srikumar Acharya

3.1 INTRODUCTION

In today's global scenario, transportation has played a vital role in the growth of the industrial sector. It has been extended to many other fields. Classically, transportation problem (TP) is regarded as a basic network problem using linear programming. The optimum solution to the classical TP requires the determination of several products to be transported from each source to various destinations, satisfying the constraints of demand and supply. Hitchcock (1941) formulated the transportation model as a simple model with sources and destinations. In a TP, a product needs to be transported from m sources to n destination, and their capacities are $c_1, c_2, c_3, \dots, c_n$ and $d_1, d_2, d_3, \dots, d_n$, respectively. The cost indulge in TP is regarded as a penalty associated with transporting a product from source "" to destination "j" and can be symbolically represented as r_{c_r}. Dantzig (1963) used the simplex method to solve the TP, and Arsham and Kahn (1989) developed a simplex-type algorithm to solve the general TP. The multi-objective TP was studied by Lee and Moore (1973). Diaz (1978, 1979) and Isermann (1979) proposed a procedure to generate all non-dominated solutions to the multi-objective linear TP. Aneja and Nair (1979) have presented a bi-criteria TP. For handling uncertainty in decision-making, Kaufmann and Gupta (1988) first examined the fuzzy TP. Using the chance-constrained technique, Bit et al. (1994) presented the multi-objective probabilistic TP. Das et al. (1999) presented the solution procedure of the multi-objective TP, where the cost coefficient of the objective function and the source and destination are considered. Parameters were expressed as an interval value by the decision-maker.

An optimization problem is defined as a fuzzy stochastic or fuzzy probabilistic programming (FPP) problem in which some or all of the parameters are represented by random variables. The uncertainty in the parameters is represented by fuzzy probability distributions. The distribution is calculated using the available observed fuzzy random data. There are two types of uncertainties: randomness, which can be incorporated by employing a probability distribution, and fuzziness, which can be represented by fuzzy numbers.

DOI: 10.1201/9781003642091-3

The concept of fuzzy random variable was first introduced by Zadeh (1968). It was further developed by Kwakernaak (1978), Kratschmer (2001), and Buckley (2004), according to different requirements of measurability. Buckley and Eslami (2004) and Buckley (2005) defined fuzzy probability using fuzzy numbers as parameters in the probability density function and probability mass function. These fuzzy numbers are initiated from the set of confidence intervals. The approach of fuzzy probability theory by Buckley (2005) is different from that of his predecessors and also comfortable from a computational point of view. The FPP problem is a mathematical programming problem in which both fuzziness and randomness are involved. In the recent past, Liu and Iwamura (1998) extended chance-constrained programming (CCP) from stochastic to fuzzy environments and also derived a technique of fuzzy simulation for the chance constraints, which are usually hard to convert to their crisp equivalents. Luhandjula (1983) and Buckley (2005) derived different methods to solve some typical FPP problems. All these techniques involve fuzziness and randomness in different scenarios. Recently, Zarghami (2010) suggested a multi-objective model to design optimum water supply and demand management under different kinds of uncertainty, where the uncertainty in satisfying some constraints is tackled by using the chance constraint approach. Abdelaziz (2012) presented a solution procedure for multi-objective stochastic problems where random variables can be in both objectives and constraint parameters. Nanda et al. (2006, 2008) discussed the deterministic equivalent of the FPP problem in the presence of normally distributed fuzzy random variables in different scenarios. The stochastic programming (SP) model was first formulated by Dantzig (1955), who suggested a two-stage programming technique that involves the conversion of SP models into their equivalent deterministic programming models. However, this technique suffers from the limitation that it does not allow any constraint to be violated even at a specific probability level. This gave rise to the concept of CCP, where constraints containing random variables are guaranteed to be satisfied with a certain probability. Charnes and Cooper (1959, 1963) developed the CCP concept.

CCP is a particular type of decision-making problem in which constraints having finite probability are violated. Luhandjula (2003) derived methodologies to solve the fuzzy SP problem in the presence of fuzzy random variables. A different approach for the SP problem with fuzzy data was described by Luhandjula and Gupta (1996). The FPP problem is a mathematical programming problem in which both fuzziness and randomness are involved. In the recent past, Liu and Iwamura (1998) extended CCP from stochastic to fuzzy environments and also derived a technique of fuzzy simulation for the chance constraints, which are usually hard to convert to their crisp equivalents. Luhandjula (1983) and Buckley (2005) derived different methods to solve some typical FPP problems. All these techniques involve fuzziness and randomness in different scenarios.

Multi-choice programming is a mathematical programming problem that allows the decision-maker to specify multiple options for a parameter. Chang

(2007) proposed a new approach to programming the multi-choice aspiration-level problem. He introduced multiplicative terms of binary variables to address the multi-choice aspiration level for each goal. Chang (2008) then replaced the binary variables' multiplicative terms with a continuous variable. Biswal and Acharya (2009a, 2009b) and Acharya and Biswal (2011) used binary variables to transform a multi-choice linear programming problem into an equivalent mathematical model. Using the concept of Chang (2007), Liao (2009) developed multi-segment goal programming. Acharya and Acharya (2012) generalized the transformation technique proposed by Biswal and Acharya (2009b). Biswal and Acharya (2011) used an interpolating polynomial approach to solve a multi-choice linear programming problem. After using the interpolation, the formulated mathematical model was a mixed integer nonlinear programming problem. Later, Acharya and Biswal (2011) investigated a probabilistic linear programming problem in which the right-hand-side parameters are multi-choice in nature and the remaining parameters are normally distributed independent random variables with known mean and variance in the probabilistic constraints. Acharya and Acharya (2013) proposed a generalized transformation rule for a multi-choice linear programming problem. Panda and Das (2014) applied the concept of Acharya and Acharya (2013) and proposed a technique for solving the multi-choice matrix game by taking only eight choices, out of which one is to be chosen. Patro et al. (2015) presented a chapter to solve the multi-choice goal programming problem by using binary variables, Vandermonde's interpolating polynomial, and linear least square approximation. In our model, we have used the least squares approximation technique to tackle the multi-choice parameter. Agrawal and Ganesh (2021) presented a solution to a stochastic TP involving multi-choice random parameters using Newton's divided difference interpolation.

3.2 BASIC PRELIMINARIES

Definition 3.1: A fuzzy number $\tilde{A}$ is a convex normalized fuzzy set $\tilde{A}$ of the real line R, with membership function $\mu A\colon R \to [0, 1]$, satisfying the following conditions.

a. There exists exactly one interval $I \in R$ such that $\mu A(x) = 1;\ x \in I{:}$.
b. The membership function μA is piecewise continuous.

Definition 3.2 (Lai and Hwang, 1992): A fuzzy number $\tilde{A} = \left(A^{(m)}; A^{(p)}; A^{(o)}\right)$ is said to be triangular if its membership function is strictly increasing in the interval $A(p); A(m)$ and strictly decreasing in $A(m); A(o)$ and $\mu A\ A(m) = 1$, where $A(m)$ is the core, $A(m) - A(p)$ is the left spread, and $A(o) - A(m)$ is the right spread of fuzzy number $\tilde{A}$.

Definition 3.3 (Buckley, 2005): α-Cut of the fuzzy number $\tilde{A}$ is the set $\{x \mid \mu A(x) \geq a\}$

for $0 < \alpha < 1$ and denoted $\tilde{A}[\alpha]$.

Definition 3.4 (Zimmermann, 1978): A fuzzy number $\tilde{A}$ is said to be positive if its membership function $\mu A(x) > 0; \forall x \geq 0$. It may be stated as follows: let $\tilde{A}[\alpha] = [A*, A*]$ be the α-cut of the fuzzy number $\tilde{A}$ for $0 < \alpha < 1$. $\tilde{A}$ is said to be positive if $A* > 0$.

Definition 3.5 (Nanda and Kar, 1992): Let $\tilde{A} = A(m); A(p); A(o)$ and $\tilde{B} = B(m); B(p); B(o)$ be two fuzzy numbers with α-cut $\tilde{A}[\alpha] = [A^*, A^*]$ and $\tilde{B}[\alpha] = [B^*, B^*]$,

respectively. Then $\tilde{A} \leq \tilde{B}$ iff $A* \leq B*$.

Definition 3.6 (Buckley and Eslami, 2004): Let $\tilde{X}$ be a continuous random variable with fuzzy parameter as fuzzy probability, then $\tilde{X}$ is said to be a continuous fuzzy random variable with a density function

$$f(x;\theta).\widetilde{P(X \leq x)} = \widetilde{\beta},\ \text{where, } 0 \leq \widetilde{\beta}, \leq 1.,\ \widetilde{\beta}, = \left(\beta^{(m)}, \beta^{(p)}, \beta^{(0)}\right).$$

$$\text{And } \beta^{(p)} \geq 0.$$

Definition 3.7 (Buckley and Eslami, 2004): Let $E = [c, d]$ be an event. Then the probability of the event E of continuous fuzzy random variable X is a fuzzy number whose alpha-cut is

$$\widetilde{P}(c \leq \tilde{X} \leq d)[\alpha] = min: \left\{\int_c^d f(x,\theta)\,dx \middle| \theta\epsilon\,\widetilde{\theta}, [\alpha] \int_{-\infty}^{\infty} f(x,\theta)dx = 1,\right.$$

$$max: \left\{\int_c^d f(x,\theta)\,dx \middle| \theta\epsilon\,\widetilde{\theta},[\alpha] \int_{-\infty}^{\infty} f(x,\theta)\,dx = 1\right\}\Bigg] = \left[\beta_*[\alpha], \beta^*[\alpha]\right.$$

Theorem 3.2: Fuzzy normal distribution: Let X be a random variable distributed normally with parameters, where the parameter is the mean and is the standard deviation of the random variable. Let the probability density function of the random variable $\tilde{X}$ be

$$F(X, \mu, \sigma) = \frac{1}{\sigma\sqrt{2\pi}} \cdot e^{\frac{-(x-\mu)^2}{2\sigma^2}}.$$

If the parameter or both are positive fuzzy numbers, then the logistic random variable X will be a fuzzy normally distributed random variable, denoted by $\tilde{X}$.

Now the fuzzy probability of the fuzzy normal random variable $\tilde{X}$ on the interval [c, d] is a fuzzy number whose alpha-cut is

$$\widetilde{P}(c \leq \tilde{X} \leq d)[\alpha] = \frac{1}{\sigma\sqrt{2\pi}} e^{\frac{-(X-\mu)^2}{2\sigma}},\ \mu\epsilon\,\mu[\alpha].$$

3.3 MATHEMATICAL MODEL

In this section, a mathematical model for the multi-choice multi-objective transportation (MCMOFST) problem involving a normal distribution is considered as follows:

$$\min: z_k = \sum_{i=1}^{m} \sum_{j=1}^{n} c_{ij}^{(k)} x_{ij},\ k = 1,2\ldots,\ K \quad (3.1)$$

Subject to

$$P\left(\sum_{j=1}^{n} x_{ij} \le a_i\right) \ge \gamma_i, \tilde{a}_i \in \left\{\tilde{a}_i^{(1)}, \tilde{a}_i^{(2)}, \ldots\ldots.., \tilde{a}_i^{(h_i)}\right\},\ i = 1,2,3\ldots.\ m \quad (3.2)$$

$$\left(\sum_{i=1}^{m} x_{ij} \ge b_j\right),\ b_j \in \left\{b_j^{(1)}, b_j^{(2)}, \ldots\ldots.., b_j^{(h_i')}\right\},\ j = 1,2,3..,n \quad (3.3)$$

$$x_{ij} > L_j > 0,\ j = 1,2,3,..n,\ i = 1,2,3, \ldots, m \quad (3.4)$$

where $0 < \gamma_i < 1$, and a_i, b_j are normal random variables, and c^k, $k = 1,2,3,\ldots,K$, is a multi-choice parameter.

Depending upon the situation, the following cases are to be considered.

Case 1: a_i is the fuzzy random variable, and μ_i, σ_i are the fuzzy numbers.

Case 2: b_j is the fuzzy random variable, and μ_i, σ_i are the fuzzy numbers.

Case 3: a_i and b_j are the fuzzy random variables, and $\mu_i, \sigma_i, \sigma_j$ are the fuzzy numbers.

We present the MCMOFST problem by considering fuzziness and randomness under one umbrella. Depending on above said situations, the following mathematical models are formulated.

Model I: The model of an MCMOFST problem, where $\tilde{a}_i \in \left\{\tilde{a}_i^{(1)}, \tilde{a}_i^{(2)}, \ldots\ldots.., \tilde{a}_i^{(m)}\right\}$, is a multi-choice fuzzy random variable (FRV) distributed according to a normal distribution with a known multi-choice mean and variance is presented as follows:

$$min: z_k = \sum_{i=1}^{m} \sum_{j=1}^{n} c_{ij}^{(k)} x_{ij}, k = 1,2,..,K. \quad (3.5)$$

Subject to

$$\left(\sum_{j=1}^{n} x_{ij} \le a_i\right),\ \tilde{a}_i \in \left\{\tilde{a}_i^{(1)}, \tilde{a}_i^{(2)}, \ldots\ldots.., \tilde{a}_i^{(h_i)}\right\},\ i = 1,2,3\ldots.\ m \quad (3.6)$$

$$P\left(\sum_{i=1}^{m} x_{ij} \ge \tilde{b}_j\right) \ge \tilde{y}_j',\ \tilde{b}_j \in \left\{\tilde{b}_j^{(1)}, \tilde{b}_j^{(2)}, \ldots\ldots.., \tilde{b}_j^{(h_i')}\right\},\ j = 1,2,3..,n \quad (3.7)$$

$$x_{ij} > L_j > 0, j = 1,2,3,..n, i = 1,2,3,\ldots,m \quad (3.8)$$

where $\tilde{\gamma}_j$ is a positive fuzzy number $0 < \gamma_i < 1$. The k parameter b_j is the deterministic constant.

Model II: The model of an MCMOFST problem, where $\tilde{b}_j \in \{\tilde{b}_j^{(1)}, \tilde{b}_j^{(2)}, \ldots\ldots.., \tilde{b}_j^{(n)}\}$ is a multi-choice FRV distributed according to a normal distribution with known multi-choice mean and variance, is presented as follows:

$$min: z_k = \sum_{i=1}^{m} \sum_{j=1}^{n} c_{ij}^{(k)} x_{ij}, k = 1, 2, \ldots, K. \tag{3.9}$$

Subject to

$$P\left(\sum_{j=1}^{n} x_{ij} \le \tilde{a}_i\right) \ge \tilde{y}_i, \text{'}\tilde{a}_i \in \left\{\tilde{a}_i^{(1)}, \tilde{a}_i^{(2)}, \ldots\ldots.., \tilde{a}_i^{(h_i)}\right\}, \tag{3.10}$$
$$i = 1, 2, 3 \ldots. m$$

$$P\left(\sum_{i=1}^{m} x_{ij} \ge \tilde{b}_j\right) \ge \tilde{y}_j, \tilde{b}_j \in \left\{\tilde{b}_j^{(1)}, \tilde{b}_j^{(2)}, \ldots\ldots.., \tilde{b}_j^{(h_i')}\right\}, \tag{3.11}$$
$$j = 1, 2, 3.., n$$

$$x_{ij} > L_j > 0, j = 1, 2, 3, .. n, i = 1, 2, 3, \ldots, m, \tag{3.12}$$

where $\tilde{\gamma}_j$ is a positive fuzzy number $0 < \gamma_i < 1$. The k parameter b_j is the deterministic constant.

Model III: The model of an MCMOFST problem, where $\tilde{a}_i \in \{\tilde{a}_i^{(1)}, \tilde{a}_i^{(2)}, \ldots\ldots.., \tilde{a}_i^{(h_i)}\}$, and $\tilde{b}_j \in \{\tilde{b}_j^{(1)}, \tilde{b}_j^{(2)}, \ldots\ldots.., \tilde{b}_j^{(n)}\}$ are multi-choice FRV distributed according to the normal distribution with known multi-choice mean and variance, is presented as follows:

$$min: z_k = \sum_{i=1}^{m} \sum_{j=1}^{n} c_{ij}^{(k)} x_{ij}, k = 1, 2, .., K. \tag{3.13}$$

Subject to

$$\tilde{P}\left(\sum_{j=1}^{n} x_{ij} \ge \tilde{a}_i\right) \ge \widetilde{y'}_i, i = 1, 2, \ldots, m, \tag{3.14}$$

$$\tilde{P}\left(\sum_{i=1}^{m} x_{ij} \ge \tilde{b}_j\right) \ge \widetilde{y'}_i, j = 1, 2, \ldots, n, \tag{3.15}$$

$$x_{ij} > L_j > 0, j = 1, 2, 3, ..n, i = 1, 2, 3, \ldots, m, \tag{3.16}$$

where $\widetilde{y'}_i$ and $\tilde{y}_i$ are fuzzy numbers and $c_{ij}^{(k)} \in R$.

3.4 SOLUTION METHODOLOGY

The solution procedure for the proposed mathematical model is completed in the following three steps.

3.4.1 Step 1: Transformation of Fuzzy Random Variable

In this step, fuzzy random parameters are converted to their crisp equivalent by using the technique followed by the chance constraint technique.

Theorem 4.1 If $\tilde{a}_i, i = 1,2,3,\ldots,m$ are independent FRVs distributed according to the normal distribution and $\tilde{\mu}_i, \tilde{\sigma}_i$ are fuzzy numbers, then

$$\left(\sum_{j=1}^{n} x_{ij} \le \tilde{a}_i\right) \ge \tilde{y}_i, \tilde{a}_i \in \left\{\tilde{a}_i^{(1)}, \tilde{a}_i^{(2)}, \ldots\ldots.., \tilde{a}_i^{(h_i)}\right\}, i = 1,2,3\ldots.m \quad (3.17)$$

is equivalent to

$$\sum_{j=1}^{n} x_{ij} \le \mu_{i*} \le k_{y_i^*,}\, \sigma_{i*}, \mu_{i*} \in \left\{\mu_{i*}^{(1)}, \mu_{i*}^{(2)}, \ldots\ldots.., \mu_{i*}^{(h_i)}\right\}, \sigma_{i*} \in \left\{\sigma_{i*}^{(1)}, \sigma_{i*}^{(2)}, \ldots\ldots.., \sigma_{i*}^{(h_i)}\right\}, i = 1,2,\ldots m \quad (3.18)$$

Proof: It is assumed that $\tilde{a}_i, i = 1,2,3,\ldots,m$ are normal FRVs with density function

$$f(a_i : \mu_i, \sigma_i) = \frac{1}{\sigma_i\sqrt{2\pi}} \cdot e^{\frac{-(a_i-\mu_i)^2}{2\sigma_i^2}}. \quad (3.19)$$

The parameter $\tilde{a}_i$ is a fuzzy random variable having a normal distribution. $\tilde{\mu}_i, \tilde{\sigma}_i$ are fuzzy numbers and whose α-cut are $\tilde{\mu}_i[\alpha] = \left[\mu_{i*}, \mu_{j*}\right]$ and $\tilde{\sigma}_i\left[\tilde{\sigma}_i\right] = \left[\sigma_{i*}, \sigma_{j*}\right]$.

Here, $\tilde{\gamma}_j$ is a positive fuzzy number $0 < \gamma_i < 1$, whose α-cut are $\tilde{\gamma}_i[\alpha] = \left[\gamma_{i*}, \gamma_{j*}\right]$.

Now, let us consider the ith constraint.

$$\tilde{P}\left(\sum_{j=1}^{n} x_{ij} \le \tilde{a}_i\right) \ge \tilde{\gamma}_i \quad (3.20)$$

The α-cut of the probabilistic constraint is expressed as

$$\tilde{P}\left(\sum_{j=1}^{n} x_{ij} \leq \tilde{a}_i\right)[\alpha]$$

$$= \tilde{P}(y_i < \tilde{a}_i)[\alpha], \text{ where } y_i = \sum_{j=1}^{n} x_{ij} = P(y_i < a_i)|a_i \epsilon\, \tilde{a}_i\, [\alpha]$$

$$= \left\{1 - P\left(\frac{a_i - \mu_i}{\sigma_i} \leq \frac{y_i - \mu_i}{\sigma_i}\right) : a_i[\alpha]\right\}$$

$$= \left\{1 - \frac{1}{\sqrt{2\pi}} \int_{-\infty}^{\frac{y_i - \mu_i}{\sigma_i}} e^{\frac{-x^2}{2}} dz : \Delta\right\}$$

$$\text{where } \Delta_1 = \{\mu_i \,\epsilon\, \tilde{\mu}_i[\alpha], \sigma_i \,\epsilon\, \tilde{\sigma}_i[\alpha]\} \text{ and } z_i = \frac{a_{i-\mu_i}}{\sigma_i}$$

$$= \left\{1 - \varphi\left(\frac{y_i - \mu_i}{\sigma_i}\right)\right\}$$

$$= \{[k_{i*}, k_i^*]\}$$

where k_{i*} is the optimal objective value of the optimization problem.

$$\min\left\{1 - \varphi\left(\frac{y_i - \mu_i}{\sigma_i}\right)\right\}$$

$$\mu_{i*} \leq \mu_i \leq \mu_i^*$$

$$\sigma_{i*} \leq \sigma_i \leq \sigma_i^*$$

where k_i^* is the optimal objective value of the optimization problem.

$$\mu_{i*} \leq \mu_i \leq \mu_i^*$$

$$\sigma_{i*} \leq \sigma_i \leq \sigma_i^*$$

$$\text{Let } \min\left[1 - \varphi\left(\frac{y_i - \mu_i}{\sigma_i}\right)\right] = 1 - \varphi\left(\frac{y_i - \mu_{i*}}{\sigma_{i*}}\right) = k_{i*}\ .$$

$$\max\left[1 - \varphi\left(\frac{y_i - \mu_i}{\sigma_i}\right)\right] = 1 - \varphi\left(\frac{y_i - \mu_i^*}{\sigma_i^*}\right) = k_i^*$$

Now we have

$$\tilde{P}\left(\sum_{j=1}^{n} x_{ij} \leq \tilde{a}_i\right)[\alpha] = 1-\varphi\left(\frac{y_i-\mu_{i^*}}{\sigma_{i^*}}\right), 1-\varphi\left(\frac{y_i-\mu_i^*}{\sigma_i^*}\right). \tag{3.21}$$

Using fuzzy inequality, the α-cut of the fuzzy constraint is expressed as

$$\begin{aligned}
&\tilde{P}\left(\sum_{j=1}^{n} x_{ij} \leq \tilde{a}_i\right)[\alpha] \geq \tilde{y}_i[\alpha] \\
&= 1-\varphi\left(\frac{y_i-\mu_{i^*}}{\sigma_{i^*}}\right) \geq \gamma_i^* \\
&= \varphi\left(\frac{y_i-\mu_{i^*}}{\sigma_{i^*}}\right) \geq 1-\gamma_i^* \\
&= \varphi\left(\frac{y_i-\mu_{i^*}}{\sigma_{i^*}}\right) \leq \varphi\left(k_{\gamma_i^*}\right) \\
&= y_i - \mu_i \leq k_{i^*}\sigma_{i^*} \\
&= y_i \leq \mu_{i^*} + k_{\gamma_i^*}\sigma_{i^*} \\
&= \sum_{j=1}^{n} x_{ij} \leq \mu_{i^*} + k_{\gamma_i^*}\sigma_{i^*}
\end{aligned}$$

Hence, the theorem is proved.

The deterministic equivalent of the MCMOFST problem (3.5–3.8) is expressed as

MOP1:

$$min\text{: } z_k = \sum_{i=1}^{m}\sum_{j=1}^{n} c_{ij}^{(k)} x_{ij}, \ k = 1,2,\ldots,K, \tag{3.22}$$

Subject to

$$\sum_{j=1}^{n} x_{ij} \leq \mu_{i^*} + k_{\gamma_i^*}\sigma_{i^*}, \tag{3.23}$$

$$\sum_{i=1}^{m} x_{ij} \geq b_j, \ j = 1,2,3,\ldots,n, \tag{3.24}$$

$$x_{ij} > L_j > 0, \ j = 1,2,3,..n. \tag{3.25}$$

Here, $\tilde{\gamma}_j$ is a positive fuzzy number and $c_j^{(k)} \in R$.

Theorem 4.2 If $\tilde{b}_j, j = 1,2,3,\ldots,n$ are independent FRVs distributed according to the normal distribution and $\tilde{\mu}_j, \tilde{\sigma}_j$ are fuzzy numbers, then

$$P\left(\sum_{i=1}^{m} x_{ij} \geq \tilde{b}_j\right) \geq \tilde{\tilde{\gamma}}'_j, \tilde{b}_j \in \left\{\tilde{b}_j^{(1)}, \tilde{b}_j^{(2)}, \ldots\ldots.., \tilde{b}_j^{(h_i')}\right\}, j = 1,2,3..,n . \quad (3.26)$$

Is equivalent to

$$\sum_{j=1}^{n} x_{ij} \leq \mu_{i*} + k_{y_i^*} \sigma_{i*}, \mu_{i*} \in \left\{\mu_{i*}^{(1)}, \mu_{i*}^{(2)}, \ldots, \mu_{i*}^{(h_i)}\right\},$$
$$\sigma_{i*} \in \left\{\sigma_{i*}^{(1)}, \sigma_{i*}^{(2)}, \ldots\ldots, \sigma_{i*}^{(h_i)}\right\}, i = 1,2, \ldots m. \quad (3.27)$$

Proof: It is assumed that $\tilde{b}_j, j = 1,2,3, \ldots, n$ are normal FRVs with density function

$$f\left(b_j : \mu_j, \sigma_j\right) = \frac{1}{\sigma_j \sqrt{2\pi}} \cdot e^{\frac{-(b_j - \mu_j)^2}{2\sigma_j^2}} . \quad (3.28)$$

The parameter $\tilde{b}_j$ is a fuzzy random variable having a normal distribution. $\tilde{\mu}_j$ and $\tilde{\sigma}_j$ are fuzzy numbers and whose α-cut are $\tilde{\mu}_j[\alpha] = \left[\mu_{i*}, \mu_i^*\right]$ and $\tilde{\sigma}_j[\alpha] = \left[\sigma_{i*}, \sigma_i^*\right]$, respectively.

Now let us consider

$$P\left(\sum_{i=1}^{m} x_{ij} \geq \tilde{b}_j\right) \geq \tilde{\gamma}'_j, \quad (3.29)$$

where $\tilde{\gamma}'_j, (i = 1,2,3, \ldots, n)$ are fuzzy numbers whose α-cut are $\tilde{\tilde{\gamma}}_j[\alpha] = \left[\tilde{\gamma}_{i*}, \tilde{\gamma}_i^*\right]$.

The α-cut of the probabilistic constraint is expressed as

$$P\left(\sum_{i=1}^{m} x_{ij} \geq \tilde{b}_j\right)[\alpha]$$
$$= \tilde{P}\left(y_i \leq \tilde{b}_j\right)[\alpha] \text{ where } y_i = \sum_{i=1}^{m} x_{ij}$$
$$= \left\{P\left(y_i \geq \tilde{b}_j\right) \middle| b_j \in \tilde{b}_j[\alpha]\right\}$$
$$= \left\{1 - P\left(\frac{b_j - \mu_j}{\sigma_j} \leq \frac{y_j - \mu_j}{\sigma_j}\right) : b_j \in \tilde{b}_j[\alpha]\right\}$$

$$= \left\{ 1 - \frac{1}{\sqrt{2\pi}} \int_{-\infty}^{\frac{y_i - \mu_j}{\sigma_j}} e^{\frac{-x^2}{2}} dz : \Delta_1 \right\}, \Delta_1 = \left\{ \mu_j \in \tilde{\mu}_j[\alpha], \sigma_j \in \tilde{\sigma}_j[\alpha] \right\} \text{ and}$$

$$z_i = \frac{b_j - \mu_j}{\sigma_j}$$

$$= 1 - \varphi\left(\frac{y_i - \mu_j}{\sigma_j} \right)$$

$$= \left\{ \left[k_{j*}, k_j^* \right] \right\}$$

where k_{j*} is the optimal objective values of the optimization problem.

$$\min \left\{ 1 - \varphi\left(\frac{y_i - \mu_j}{\sigma_j} \right) \right\}$$

$$\mu_{j*} \le \mu_j \le \mu_j^*,$$

$$\sigma_{j*} \le \sigma_j \le \sigma_j^*$$

where k_j^* is the optimal objective values of the optimization problem.

$$\mu_{j*} \le \mu_j \le \mu_j^*$$

$$\sigma_{j*} \le \sigma_j \le \sigma_j^*$$

$$\text{Let } \min\left[1 - \varphi\left(\frac{y_i - \mu_j}{\sigma_i} \right) \right] = 1 - \varphi\left(\frac{y_j - \mu_{j*}}{\sigma_{j*}} \right) = k_{j*}$$

$$\max\left[1 - \varphi\left(\frac{y_i - \mu_j}{\sigma_j} \right) \right] = 1 - \varphi\left(\frac{y_i - \mu_j^*}{\sigma_j^*} \right) = k_j^*$$

Now we have

$$\tilde{P}\left(\sum_{j=1}^{n} x_{ij} \le \tilde{b}_i \right)[\alpha] = 1 - \varphi\left(\frac{y_i - \mu_{j*}}{\sigma_{j*}} \right), 1 - \varphi\left(\frac{y_i - \mu_j^*}{\sigma_j^*} \right). \tag{3.30}$$

Using fuzzy inequality, the α-cut of the fuzzy constraint (3.28) is expressed as

$$\tilde{P}\left(\sum_{j=1}^{n} x_{ij} \leq \tilde{a}_i\right)[\alpha] \geq \tilde{y}_i[\alpha]$$

$$= 1 - \varphi\left(\frac{y_i - \mu_{i^*}}{\sigma_{i^*}}\right) \geq \gamma_i^*$$

$$= \varphi\left(\frac{y_i - \mu_{i^*}}{\sigma_{i^*}}\right) \geq 1 - \gamma_i^*$$

$$= \varphi\left(\frac{y_i - \mu_{i^*}}{\sigma_{i^*}}\right) \leq \varphi\left(k_{\gamma_i^*}\right)$$

$$= \left(\frac{y_i - \mu_{i^*}}{\sigma_{i^*}}\right) \leq k_{\gamma_i^*}$$

$$= y_i - \mu_{i^*} \leq \sigma_{i^*} k_{\gamma_i^*}$$

$$= \sum_{j=1}^{n} x_{ij} \leq \mu_{i^*} + k_{\gamma_i^*} \sigma_{i^*}$$

Hence the theorem is proved.

The deterministic equivalent of the MCMOFST problem (3.9–3.11) is expressed as

MOP2:

$$min: z_k = \sum_{i=1}^{m} \sum_{j=1}^{n} c_{ij}^{(k)} x_{ij},\ k = 1,2,\ldots,K. \tag{3.31}$$

Subject to

$$\sum_{j=1}^{n} x_{ij} \leq a_i,\ i = 1,2,3,..\,n, \tag{3.32}$$

$$\sum_{i=1}^{m} x_{ij} \geq \mu_j^* + \sigma_{i^*} k_{\gamma_i^*},\ j = 1,2,3,\ldots,n, \tag{3.33}$$

$$x_{ij} > L_j > 0,\ j = 1,2,3,..\,n, \tag{3.34}$$

where $\tilde{\gamma}_j$ is a positive fuzzy number and $c_j^{(k)} \in R$.

Theorem 4.3 When both $\tilde{a}_i, i = 1,2,3,\ldots,m$ and $\tilde{b}_j, j = 1,2,3,\ldots,n$ are independent FRVs distributed according to normal distribution. In this case, the equivalent deterministic model for the MCMOFST problem is already proved in Theorems 4.1 and 4.2.

The deterministic equivalent of the MCMOFST problem (3.13–3.16) is expressed as

MOP3:

$$\min: z_k = \sum_{i=1}^{m} \sum_{j=1}^{n} c_{ij}^{(k)} x_{ij},\ k = 1,2,3 \ldots, K. \tag{3.35}$$

Subject to

$$\sum_{j=1}^{n} x_{ij} \leq \mu_{i*} + k_{\gamma_i'} \sigma_{i*},\ j = 1,2 \ldots, n, \tag{3.36}$$

$$\sum_{i=1}^{m} x_{ij} \leq \mu_{i*} + k_{\gamma_i'} \sigma_{i*},\ j = 1,2 \ldots, n. \tag{3.37}$$

$$x_{ij} > L_j > 0,\ j = 1,2,3, .. n, \tag{3.38}$$

where $\tilde{\gamma}_j$ is fuzzy number and $c_j^{(k)} \in R$.

3.4.2 Step 2: Dealing with Multi-Choice Parameters

In this section, we consider a power function, a polynomial of the nth degree to the given data points. (Fitting of nth degree polynomial when $(\mathrm{n}+1)$ points are given).

0	1	2	3	. . .	$n-2$	$n-1$
f (0)	F (1)	f (2)	f (3)	. . .	f($n-2$)	f($n-1$)

Polynomial of the nth degree: A polynomial that passes through a given set of points is called an interpolating polynomial. There are n points in a plane (i, y_i), $i = 1,2, \ldots, n$ with distinct i, then there exists a unique polynomial in x of degree $(n-1)$. Let the polynomial be,

$$\begin{aligned} P_{Rk} - 1^{(z^{(i)})} &= a_0 + a_1 z^i + a_2 z^{i2} + a_3 z^{i3} + a_4 z^{i4} +, \ldots. \\ &\quad + a_{n-2} z^{in-2} + a_{n-1} z^{in-1}. \end{aligned}$$

After substituting the "n" points in the polynomial, we have

$$\begin{aligned} f(0) = P(0) &= a_0 + a_1(0) + a_2(0)^2 + a_3(0)^3 + a_4(0)^4 + \ldots\ldots\ldots\ldots + \\ &\quad a_{n-2}(0)^{n-2} + a_{n-1}(0)^{n-1} = y_0, \\ f(1) = P(1) &= a_0 + a_1(1) + a_2(1)^2 + a_3(1)^3 + a_4(1)^4 + \ldots\ldots\ldots\ldots + \\ &\quad a_{n\text{-}2}(1)^{n\text{-}2} + a_{\mathrm{n}\text{-}1}(1)^{n\text{-}1} = y_1, \\ f(2) = P(2) &= a_0 + a_1(2) + a_2(2)^2 + a_3(2)^3 + a_4(2)^4 + \ldots\ldots\ldots\ldots\ldots\ldots + \\ &\quad a_{n-2}(2)^{n-2} + a_{n-1}(2)^{n-1} = y_2, \end{aligned}$$

$$f(n-1) = P(n-1) = a_0 + a_1(n-1) + a_2(n-1)^2 + a_3(n-1)^3 + a_4(n-1)^4 + \ldots + a_{n-2}(n-1)^{n-2} + a_{n-1}(n-1)^{n-1} = y_{n-1}.$$

The system of equations can be written as $AX = B$, where

$$A = \begin{bmatrix} 1 & 0 & 0 & 0 & 0 & \ldots & 0 & 0 \\ 1 & 1 & 1^2 & 1^3 & 1^4 & \ldots & 1^{n-2} & 1^{n-1} \\ 1 & 2 & 2^2 & 2^3 & 2^4 & \ldots & 2^{n-2} & 2^{n-1} \\ \vdots & \vdots & \vdots & \vdots & \vdots & \ddots & \vdots & \vdots \\ 1 & n-1 & n-1^2 & n-1^3 & n-1^4 & \ldots & n-1^{n-2} & n-1^{n-1} \end{bmatrix},$$

$$X = \begin{bmatrix} a_0 \\ a_1 \\ a_2 \\ \vdots \\ a_{n-2} \end{bmatrix},$$

$$Y = \begin{bmatrix} y_0 \\ y_1 \\ y_2 \\ \vdots \\ y_{n-2} \end{bmatrix}.$$

The value of $a_1, a_2, \cdots a_{n-1}$ can be found out by solving $AX = B$. If A is non-singular, then the system can be solved. Now by using interpolating polynomial, the model (3.34–3.37) is expressed as

$$\min : z_k = \sum_{i=1}^{m} \sum_{j=1}^{n} c_{ij}^{(k)} x_{ij},\ k = 1,2,3 \ldots, K. \tag{3.39}$$

Subject to

$$\sum_{j=1}^{n} x_{ij} \le P_{(Ma_i-1)}(z),\ i = 1,2,3,..\, m, \tag{3.40}$$

$$\sum_{j=1}^{n} x_{ij} \le P_{(Nb_j-1)}(z),\ j = 1,2,3,..\, m, \tag{3.41}$$

$$x_{ij} > L_j > 0,\ j = 1,2,3,..\, n, \tag{3.42}$$

where $P_{(Ma_i-1)}$ and $P_{(Nb_j-1)}$ are polynomial in Z with order.

$$P_{Rk} - 1^{\left(z^{(i)}\right)} = a_0 + a_1 z^i + a_2 z^{i2} + a_3 z^{i3} + a_4 z^{i4} +, \ldots, + a_{n-2} z^{in-2} + a_{n-1} z^{in-1},$$

where z is an integer, and $0 \le z \le Ma_i - 1$ and $0 \le z \le Nb_j - 1$. F is a feasible region of the transformed constraint.

3.4.3 Step 3: Fuzzy Programming Method

Steps of the fuzzy programming technique are given below:

Step 1: Select the first objective function (i.e., Zk(x), k = 1) and solve it as a single objective mathematical programming (MP) problem subject to the constraints. Let x (1) be the ideal solution. Then select the second objective function and find the ideal solution as x(2), continue the process K times for K different objective functions. Let x(1), x(2), ..., x(K) be the ideal solutions for the respective objective functions.

Step 2: Evaluate all these objective functions at all these ideal solutions and formulate a pay-off matrix (Table 3.1) of size by K as follows.

Step 3: From pay-off matrix (Table 3.1) determine the bounds for the k-th objective function $Z_{k(x)}$, $r = 1,2,3,\ldots,k$. If an objective function is of the maximization type, find the best upper bound U_k^* and worst lower bound L_K^-.

If an objective function is of the minimization type, find the lower bound L_K^* and worst upper bound U_K^-, $k = 1,2,3, \ldots, K$.

Step 4: Associate a membership function $\mu Z_{k(x)}$ to the k-th objective function $Z_{k(x)}$.

$$\mu_{Z_k}(x) = \begin{cases} 1 & if\ Z_k(x) \ge U_k^* \\ \dfrac{Z_k(x) - L_k^-}{U_k^* - L_k^-}, & if\ L_k^- < Z_k(x) < U_k^*, \quad k = 1,2,3 \ldots, K \\ 0 & if\ Z_k(x) \le L_k^- \end{cases} \tag{3.43}$$

$$\mu_{Z_k}(x) = \begin{cases} 1 & if\ Z_k(x) \le L_k^* \\ \dfrac{U_k^- - Z_k(x)}{U_k^- - L_k^*}, & if\ L_k^* < Z_k(x) < U_k^-, \quad k = 1,2,3 \ldots, K \\ 0 & if\ Z_k(x) \le U_k^- \end{cases} \tag{3.44}$$

Table 3.1 Pay-Off Matrix.

	$Z_1(x)$	$Z_2(x)$	...	$Z_K(x)$
x(1)	Z_{11}	Z_{12}	...	Z_{1K}
x(2)	Z_{21}	Z_{22}	...	Z_{2K}
.	.	.	.	.
x(K)	Z_{K1}	Z_{K2}	...	Z_{KK}

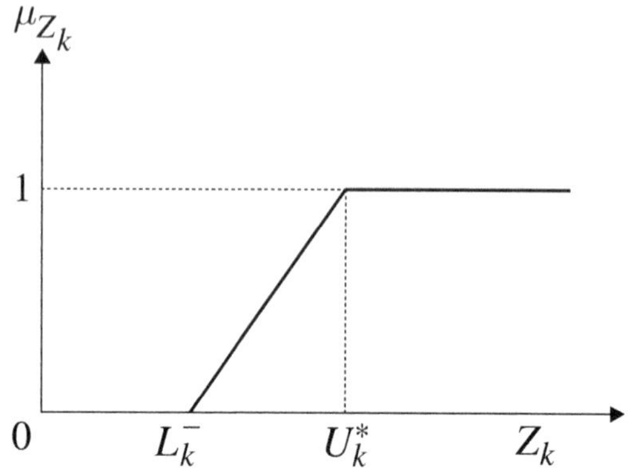

Figure 3.1 Membership function of a vector maximization problem.

Step 5(a): Use a max–min operator with an augmented variable and formulate a single objective crisp MP problem as

$$max : \lambda. \tag{3.45}$$

Subject to

$$\lambda \leq \mu Z_{k(x)}, K = 1, 2, 3, \dots, K, \tag{3.46}$$

$$x \in S, \tag{3.47}$$

where S is the feasible region of the multi objective mathematical progrmming (MOMP) model.

Step 5(b): Similarly, if we use a min–max operator with an augmented variable, a single objective crisp MP problem can be formulated as

$$min : \lambda. \tag{3.48}$$

Subject to

$$\lambda \geq \mu Z_{k(x)}, K = 1, 2, 3, \dots, K, \tag{3.49}$$

$$x \in S, \tag{3.50}$$

where S is the feasible region of the MOMP model.

Step 6: Solve the crisp model by using an appropriate mathematical programming method to an optimal compromise solution x^*. Then evaluate all the objective functions at the optimal compromise solution x^*.

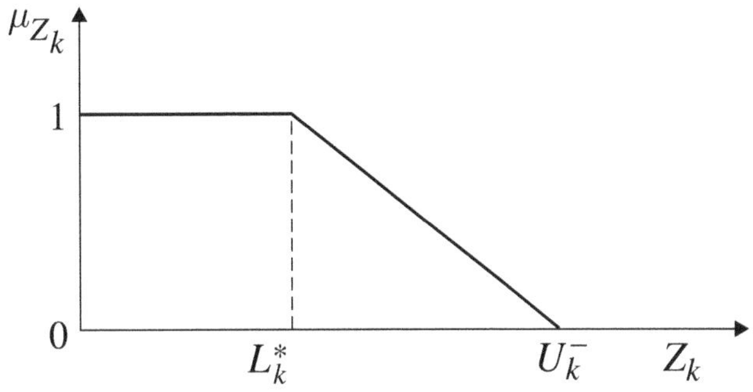

Figure 3.2 Membership function of a vector minimization problem.

3.5 NUMERICAL EXAMPLE

Consider the following example:

$$\begin{aligned}\min: Z_1(x) = {} & x_{11} + 2x_{12} + 7x_{13} + 7x_{14} + x_{21} + 9x_{22} \\ & + 3x_{23} + 4x_{24} + 8x_{31} + 9x_{32} + 4x_{33} + 6x_{34},\end{aligned} \tag{3.51}$$

$$\begin{aligned}\min: Z_2(x) = {} & 4x_{11} + 4x_{12} + 3x_{13} + 3x_{14} + 5x_{21} + 8x_{22} \\ & + 9x_{23} + 10x_{24} + 6x_{31} + 2x_{32} + 5x_{33} + x_{34}.\end{aligned} \tag{3.52}$$

Subject to

$$\tilde{P}\left(x_{11} + x_{12} + x_{13} + x_{14} \preceq \widetilde{a_1}\right) \succeq \widetilde{0.4} \Big| a_1 \epsilon \left\{\widetilde{a_{11}}, \widetilde{a_{12}}, \widetilde{a_{13}}\right\}, \tag{3.53}$$

$$\tilde{P}\left(x_{21} + x_{22} + x_{23} + x_{24} \preceq \widetilde{a_2}\right) \succeq \widetilde{0.5} \Big| a_2 \epsilon \left\{\widetilde{a_{21}}, \widetilde{a_{22}}, \widetilde{a_{23}}\right\}, \tag{3.54}$$

$$\tilde{P}\left(x_{31} + x_{32} + x_{33} + x_{34} \preceq \widetilde{a_3}\right) \succeq \widetilde{0.4} \Big| a_3 \epsilon \left\{\widetilde{a_{31}}, \widetilde{a_{32}}, \widetilde{a_{33}}\right\}, \tag{3.55}$$

$$\tilde{P}\left(x_{11} + x_{21} + x_{31} \succeq \widetilde{b_1}\right) \succeq \widetilde{0.8} \Big| b_1 \epsilon \left\{\widetilde{b_{11}}, \widetilde{b_{12}}, \widetilde{b_{13}}\right\}, \tag{3.56}$$

$$\tilde{P}\left(x_{12} + x_{22} + x_{32} \succeq \widetilde{b_2}\right) \succeq \widetilde{0.8} \Big| b_2 \epsilon \left\{\widetilde{b_{21}}, \widetilde{b_{22}}, \widetilde{b_{23}}\right\}, \tag{3.57}$$

$$\tilde{P}\left(x_{13} + x_{23} + x_{33} \succeq \widetilde{b_3}\right) \succeq \widetilde{0.7} \Big| b_3 \epsilon \left\{\widetilde{b_{31}}, \widetilde{b_{32}}, \widetilde{b_{33}}\right\}, \tag{3.58}$$

$$\tilde{P}\left(x_{14} + x_{24} + x_{34} \succeq \widetilde{b_4}\right) \succeq \widetilde{0.7} \Big| b_4 \epsilon \left\{\widetilde{b_{41}}, \widetilde{b_{42}}, \widetilde{b_{43}}\right\}, \tag{3.59}$$

where a_i and b_j follow fuzzy normal distribution, $i = 1,2,3,..m$ and $j = 1,2,3,..n$.

$$\tilde{\mu}_{a_{11}} = \{6,7,8\}, \tilde{\mu}_{a_{12}} = \{7,8,9\}, \tilde{\mu}_{a_{13}} = \{8,9,10\}, \tilde{\mu}_{a_{21}} = \{17,18,19\},$$
$$\tilde{\mu}_{a_{22}} = \{18,19,20\}$$

$$\tilde{\mu}_{a_{23}} = \{19,20,21\}, \tilde{\mu}_{a_{31}} = \{15,16,17\}, \tilde{\mu}_{a_{32}} = \{16,17,18\},$$
$$\tilde{\mu}_{33} = \{17,18,19\}$$

$$\tilde{\mu}_{b_{11}} = \{9,10,11\}, \tilde{\mu}_{b_{12}} = \{10,11,12\}, \tilde{\mu}_{b_{13}} = \{11,12,13\}, \tilde{\mu}_{b_{21}} = \{1,2,3\}$$

$$\tilde{\mu}_{b_{22}} = \{2,3,4\}, \tilde{\mu}_{b_{23}} = \{3,4,5\}, \tilde{\mu}_{b_{31}} = \{13,14,15\}, \tilde{\mu}_{b_{33}} = \{14,15,16\}$$

$$\tilde{\mu}_{b_{41}} = \{14,15,16\}, \tilde{\mu}_{b_{42}} = \{15,16,17\}, \tilde{\mu}_{b_{43}} = \{16,17,18\}$$

$$\tilde{\sigma}_{a_{11}} = \{2,3,4\}, \tilde{\sigma}_{a_{12}} = \{3,4,5\}, \tilde{\sigma}_{a_{13}} = \{4,5,6\}, \tilde{\sigma}_{a_{21}} = \{5,6,7\},$$
$$\tilde{\sigma}_{a_{22}} = \{6,7,8\}$$

$$\tilde{\sigma}_{a_{23}} = \{7,8,9\}, \tilde{\sigma}_{a_{31}} = \{5,6,7\}, \tilde{\sigma}_{a_{32}} = \{6,7,8\}, \tilde{\sigma}_{a_{33}} = \{7,8,9\},$$
$$\tilde{\sigma}_{b_{11}} = \{2,3,4\}$$

$$\tilde{\sigma}_{b_{12}} = \{3,4,5\}, \tilde{\sigma}_{b_{13}} = \{4,5,6\}, \tilde{\sigma}_{b_{21}} = \{1,2,3\}, \tilde{\sigma}_{b_{22}} = \{1,2,3\},$$
$$\tilde{\sigma}_{b_{23}} = \{2,3,4\}$$

$$\tilde{\sigma}_{b_{31}} = \{3,4,5\}, \tilde{\sigma}_{a_{12}} = \{4,5,6\}, \tilde{\sigma}_{a_{33}} = \{5,6,7\}$$

$$\tilde{\sigma}_{b_{41}} = \{4,5,6\}, \tilde{\sigma}_{b_{42}} = \{5,6,7\}, \tilde{\sigma}_{b_{43}} = \{6,7,8\}$$

Solution:

Using Theorem 4.1, the deterministic equivalent model is

$$\begin{aligned} \min: Z_1(x) = {} & x_{11} + 2x_{12} + 7x_{13} + 7x_{14} + x_{21} + 9x_{22} \\ & + 3x_{23} + 4x_{24} + 8x_{31} + 9x_{32} + 4x_{33} + 6x_{34}, \end{aligned} \tag{3.60}$$

$$\begin{aligned} \min: Z_2(x) = {} & 4x_{11} + 4x_{12} + 3x_{13} + 3x_{14} + 5x_{21} + 8x_{22} \\ & + 9x_{23} + 10x_{24} + 6x_{31} + 2x_{32} + 5x_{33} + x_{34}. \end{aligned} \tag{3.61}$$

Subject to

$$x_{11} + x_{12} + x_{13} + x_{14} \le \tilde{\mu}_{a_{1*}} + \tilde{k}_{\gamma_{1*}} \tilde{\sigma}_{a_{1*}} \left| \widetilde{a_1} \in \left\{ \widetilde{a_{11}}, \widetilde{a_{12}}, \widetilde{a_{13}} \right\} \right., \tag{3.62}$$

$$x_{21} + x_{22} + x_{23} + x_{24} \le \tilde{\mu}_{a_{2*}} + \tilde{k}_{\gamma_{2*}} \tilde{\sigma}_{a_{2*}} \left| \widetilde{a_2} \in \left\{ \widetilde{a_{21}}, \widetilde{a_{22}}, \widetilde{a_{23}} \right\} \right., \tag{3.63}$$

$$x_{31} + x_{32} + x_{33} + x_{34} \le \tilde{\mu}_{a_{3*}} + \tilde{k}_{\gamma_{3*}} \tilde{\sigma}_{a_{3*}} \left| \widetilde{a_3} \in \left\{ \widetilde{a_{31}}, \widetilde{a_{32}}, \widetilde{a_{33}} \right\} \right., \tag{3.64}$$

$$x_{11} + x_{21} + x_{31} \ge \tilde{\mu}_{b_{1*}} + \tilde{k}_{\delta_{1*}} \tilde{\sigma}_{b_{1*}} \left| \widetilde{b_1} \in \left\{ \widetilde{b_{11}}, \widetilde{b_{12}}, \widetilde{b_{13}} \right\} \right., \tag{3.65}$$

$$x_{12} + x_{22} + x_{32} \ge \tilde{\mu}_{b_{2*}} + \tilde{k}_{\delta_{2*}} \tilde{\sigma}_{b_{2*}} \left| \widetilde{b_2} \in \left\{ \widetilde{b_{21}}, \widetilde{b_{22}}, \widetilde{b_{23}} \right\} \right., \tag{3.66}$$

$$x_{13} + x_{23} + x_{33} \geq \tilde{\mu}_{b_{3*}} + \tilde{k}_{\delta_{3*}} \tilde{\sigma}_{b_{3*}} \left| \widetilde{b_3} \in \left\{ \widetilde{b_{31}}, \widetilde{b_{32}}, \widetilde{b_{33}} \right\} \right., \tag{3.67}$$

$$x_{14} + x_{24} + x_{34} \geq \tilde{\mu}_{b_{4*}} + \tilde{k}_{\delta_{4*}} \tilde{\sigma}_{b_{4*}} \left| \widetilde{b_4} \in \left\{ \widetilde{b_{41}}, \widetilde{b_{42}}, \widetilde{b_{43}} \right\} \right., \tag{3.68}$$

$$\sum_{i=1}^{3} x_{ij} \geq 0,\ i = 1,2,3,4, \tag{3.69}$$

where

$$\tilde{\gamma}_1 = \widetilde{0.4},\ \tilde{\gamma}_2 = \widetilde{0.5},\ \tilde{\gamma}_3 = \widetilde{0.4},\ \tilde{\sigma}_1 = \widetilde{0.8},\ \tilde{\sigma}_2 = \widetilde{0.8},\ \tilde{\sigma}_3 = \widetilde{0.7},\ \tilde{\sigma}_4 = \widetilde{0.7}.$$

For $\alpha = 0.1$ the above deterministic equivalent becomes

$$\begin{aligned} \min: Z_1(x) = {} & x_{11} + 2x_{12} + 7x_{13} + 7x_{14} + x_{21} + 9x_{22} \\ & + 3x_{23} + 4x_{24} + 8x_{31} + 9x_{32} + 4x_{33} + 6x_{34}, \end{aligned} \tag{3.70}$$

$$\begin{aligned} \min: Z_2(x) = {} & 4x_{11} + 4x_{12} + 3x_{13} + 3x_{14} + 5x_{21} + 8x_{22} \\ & + 9x_{23} + 10x_{24} + 6x_{31} + 2x_{32} + 5x_{33} + x_{34}. \end{aligned} \tag{3.71}$$

Subject to

$$x_{11} + x_{12} + x_{13} + x_{14} \leq (p_1 + 6.1) + 0.279p_1 + 2.1, \tag{3.72}$$

$$x_{21} + x_{22} + x_{23} + x_{24} \leq (p_2 + 17.1) + 0.025p_2 + 5.1, \tag{3.73}$$

$$x_{31} + x_{32} + x_{33} + x_{34} \leq (p_3 + 15.1) + 0.279p_2 + 3.1, \tag{3.74}$$

$$x_{11} + x_{21} + x_{31} \geq (q_1 + 9.1) + 0.806q_1 + 2.1, \tag{3.75}$$

$$x_{12} + x_{22} + x_{32} \geq (q_2 + 1.1) + 0.806q_2 + 1.1, \tag{3.76}$$

$$x_{13} + x_{23} + x_{33} \geq (q_3 + 12.1) + 0.496q_3 + 3.1, \tag{3.77}$$

$$x_{14} + x_{24} + x_{34} \geq (q_4 + 14.1) + 0.496q_4 + 4.1, \tag{3.78}$$

$$\sum_{i=1}^{3} x_{ij} \geq 0,\ j = 1,2,3,4. \tag{3.79}$$

The above deterministic multi-objective multi-choice fuzzy stochastic programming problem is solved using the fuzzy programming method. The two ideal solutions obtained are

$$\begin{aligned} X^1 &= \begin{pmatrix} x_{11}, x_{12}, x_{13}, x_{14}, x_{21}, x_{22}, x_{23}, x_{24}, x_{31}, x_{32}, x_{33}, x_{34}, x_{41}, \\ x_{42}, x_{43}, x_{44} \end{pmatrix} \\ &= (7.257300, 1.986600, 0, 0, 3.535300, 0, 0, 15.7422, 0, 0, \\ &\quad 13.63760, 0.3914000) \end{aligned}$$

$$p_1 = 2, p_2 = 2, p_3 = 2, q_1 = 0, q_2 = 0, q_3 = 0, q_4 = 0; Z_1 = 134.6334.$$

$$X^2 = \begin{pmatrix} x_{11}, x_{12}, x_{13}, x_{14}, x_{21}, x_{22}, x_{23}, x_{24}, x_{31}, x_{32}, x_{33}, x_{34}, x_{41}, \\ x_{42}, x_{43}, x_{44} \end{pmatrix}$$

$$= \begin{pmatrix} 0, 0, 9.243900, 0, 10.79260, 0, 3.433000, 0, 0, 1.986600, \\ 0.9667000, 16.13360 \end{pmatrix};$$

$$p_1 = 2, p_2 = 2, p_3 = 2, q_1 = 0, q_2 = 0, q_3 = 0, q_4 = 0; Z_2 = 137.5020.$$

The pay-off matrix formulated as:

Now applying the min–max operator, the final form of the problem is

$$\min : \lambda, \tag{3.80}$$

$$\text{Subject to} : x_{11} + 2x_{12} + 7x_{13} + 7x_{14} + x_{21} + 9x_{22} + 3x_{23} + 4x_{24} + 8x_{31} + 9x_{32} + 4x_{33} + 6x_{34} - \lambda 105.1073 \geq 134.6334, \tag{3.81}$$

$$4x_{11} + 4x_{12} + 3x_{13} + 3x_{14} + 5x_{21} + 8x_{22} + 9x_{23} + 10x_{24} + 6x_{31} + 2x_{32} + 5x_{33} + x_{34} - \lambda 66.846 \geq 105.1073. \tag{3.82}$$

Table 3.2 Pay-Off Matrix.

	Z_1	Z_2
X_1	134.6334	204.3467
X_2	239.7407	137.5020

Subject to

$$x_{11} + x_{12} + x_{13} + x_{14} \leq (p_1 + 6.1) + 0.279p_1 + 2.1, \tag{3.83}$$

$$x_{21} + x_{22} + x_{23} + x_{24} \leq (p_2 + 17.1) + 0.025p_2 + 5.1, \tag{3.84}$$

$$x_{31} + x_{32} + x_{33} + x_{34} \leq (p_3 + 15.1) + 0.279p_2 + 3.1, \tag{3.85}$$

$$x_{11} + x_{21} + x_{31} \geq (q_1 + 9.1) + 0.806q_1 + 2.1, \tag{3.86}$$

$$x_{12} + x_{22} + x_{32} \geq (q_2 + 1.1) + 0.806q_2 + 1.1, \tag{3.87}$$

$$x_{13} + x_{23} + x_{33} \geq (q_3 + 12.1) + 0.496q_3 + 3.1, \tag{3.88}$$

$$x_{14} + x_{24} + x_{34} \geq (q_4 + 14.1) + 0.496q_4 + 4.1, \tag{3.89}$$

$$\sum_{i=1}^{3} x_{ij} \geq 0, j = 1,2,3,4. \tag{3.90}$$

The crisp problem is solved using Lingo [16] software, and an optimal compromise solutions obtained for different values are given in Table 3.3.

Table 3.3 Solutions for Different Values of α .

α	X_{ij} $(i=1,2,3,4; j=1,2,$	λ	Z_1	Z_2	p_1, p_2, t	$q_1, q_2, q_3,$
0.1	(0, 1.151133, 8.092767, 0, 10.79260, 0, 0, 0.8354674, 2.111833, 16.13360)	0.5535103	192.8114	142.1065	2, 2, 2	0, 0, 0, 0
0.2	(0, 1.384106, 8.096894, 0, 10.89840, 0, 3.112600, 0, 0, 0.7422944, 2.488106, 16.16560)	0.558248	193.3093	142.4232	2, 2, 2	0, 0, 0, 0
0.3	(0, 1.637876, 8.088985, 0, 10.99970, 0, 2.753500, 0, 0, 0.6223931, 2.908905, 16.19320)	0.5600166	193.4093	142.5415	2, 2, 2	0, 0, 0, 0
0.4	(0, 1.899413, 8.075787, 0, 11.09440, 0, 2.377600, 0, 0, 0.4889531, 3.34713, 16.21280)	0.5612405	193.6339	142.6233	2, 2, 2	0, 0, 0, 0
0.5	(0, 1.987788, 8.123120, 0, 11.18500, 0, 2.295000, 0, 0, 0.5232120, 3.429288, 16.23250)	0.5717559	194.7291	143.3262	2, 2, 2	0, 0, 0, 0
0.6	(0, 2.471398, 8.023852, 0, 11.25800, 0, 1.578500, 0, 0, 0.1075416, 3.430948, 16.24680)	0.5672134	193.2011	142.3514	2, 2, 2	0, 0, 0, 0
0.7	(0.05401712, 2.742100, 7.971883, 0, 11.30186, 0, 1.128917, 0, 0, 0.4827600, 16.26040)	0.5543728	192.902	142.1642	2, 2, 2	0, 0, 0, 0
0.8	(0.3573493, 2.849400, 7.839651, 0, 11.07505, 0, 0.9329493, 0, 0, 5.186400, 16.26400)	0.5566103	193.1372	142.3137	2, 2, 2	0, 0, 0, 0
0.9	(0.6814321, 2.950700, 7.692368, 0, 10.51927, 0, 0.7415321, 0, 0, 0.551300, 16.26710)	0.5582007	193.3044	142.4201	2, 2, 2	0, 0, 0, 0

3.6 CONCLUSIONS

The purpose of the chapter was to present a solution procedure for a multi-choice multi-objective fuzzy stochastic problem. The parameter, which is an independent fuzzy random variable, follows a multi-fuzzy normal distribution. The fuzzy stochastic model has been converted into deterministic equivalent models. In the first step, using the concept of cut, the fuzziness is removed, and in the second step, the multi-choice parameters are handled with the least square approximation technique. The third step deals with the randomness; in order to remove the randomness, the chance constraint method has been used. Finally, the fuzzy programming method is applied to solve the multi-choice multi-objective equivalent crisp mathematical model to obtain the optimal solution. The methodology has been illustrated in a numerical example. In this chapter, a fuzzy multi-choice parameter is used, and the procedure is illustrated to handle such choices. For various values, the result is obtained, and the best result is obtained for $\alpha = 01$.

Acknowledgement: The authors wish to thank the reviewers for their valuable comments that helped increase the quality and clarity of the chapter.

REFERENCES

Abdelaziz, F. (2012). Solution approaches for the multiobjective stochastic programming, *European Journal of Operational Research* 216(1): 1–16.

Acharya, S. and Acharya, M. M. (2012). Generalized transformation techniques for multi-choice linear programming problems, *An International Journal of Optimization and Control: Theories & Applications (IJOCTA)* 3(1): 4554.

Acharya, S. and Acharya, M. M. (2013). Generalized transformation techniques for multi-choice linear programming problems, *International Journal of Optimization & Control: Theories & Applications* 3(1): 45–54.

Acharya, S. and Biswal, M. P. (2011). Solving probabilistic programming problems involving multi-choice parameters, *Opsearch* 48(3): 217–235.

Acharya, S., Ranarahu, N., Dash, J. K. and Acharya, M. M. (2014). Solving multi-objective fuzzy probabilistic programming problem, *Journal of Intelligent & Fuzzy Systems* 26(2): 935–948.

Agrawal, P. and Ganesh, T. (2021). Solution of stochastic transportation problem involving multi-choice random parameter using Newton's divided difference interpolation, *Journal of Information and Optimization Sciences* 42(1): 77–91.

Aneja, Y. P. and Nair, K. (1979). Bicriteria transportation problem, *Management Science* 25(1): 73–78.

Arsham, H. and Kahn, A. (1989). A simplex-type algorithm for general transportation problems: An alternative to stepping-stone, *Journal of the Operational Research Society*: 581–590.

Biswal, M. and Acharya, S. (2009a). Multi-choice multi-objective linear programming problem, *Journal of Interdisciplinary Mathematics* 12(5): 606–637.

Biswal, M. and Acharya, S. (2009b). Transformation of a multi-choice linear programming problem, *Applied Mathematics and Computation* 210(1): 182–188.

Biswal, M. and Acharya, S. (2011). Solving multi-choice linear programming problems by interpolating polynomials, *Mathematical and Computer Modelling* 54(5): 1405–1412.

Bit, A., Biswal, M. and Alam, S. (1994). Fuzzy programming approach to chance-constrained multi-objective transportation problem, *The Journal of Fuzzy Mathematics* 2: 117–130.

Buckley, J. (2004). Uncertain probabilities III: the continuous case, *Soft Computing—A Fusion of Foundations, Methodologies and Applications* 8(3): 200–206.

Buckley, J. (2005). *Fuzzy probabilities: new approach and applications*, Vol. 115. Springer Verlag, Berlin.

Buckley, J. and Eslami, E. (2004). Uncertain probabilities II: the continuous case, *Soft Computing—A Fusion of Foundations, Methodologies and Applications* 8(3): 193–199.

Chang, C. (2007). Multi-choice goal programming, *Omega* 35(4): 389–396.

Chang, C.-T. (2008). Revised multi-choice goal programming, *Applied Mathematical Modelling* 32(12): 2587–2595.

Charnes, A. and Cooper, W. (1959). Chance-constrained programming, *Management Science* pp. 73–79.

Charnes, A. and Cooper, W. (1963). Deterministic equivalents for optimizing and satisficing under chance constraints, *Operations Research* pp. 18–39.

Dantzig, G. (1955). Linear programming under uncertainty, *Management Science* 1(3-4): 197–206.

Dantzig, G. (1963). *Linear programming and extensions*. Princeton University Press, Princeton.

Das, S., Goswami, A. and Alam, S. (1999). Multiobjective transportation problem with interval cost, source and destination parameters, *European Journal of Operational Research* 117(1): 100–112.

Diaz, J. (1978). Solving multiobjective transportation problems, *Ekonomicko-Matematicky Obzor* 14(3): 267–274.

Diaz, J. (1979). Finding a complete description of all efficient solutions to a multiobjective transportation problem, *Ekonomicko-Matematicky Obzor* 15(1): 62–73.

Hitchcock, F. L. (1941). The distribution of a product from several sources to numerous localities, *J. Math. Phys* 20(2): 224–230.

Isermann, H. (1979). The enumeration of all efficient solutions for a linear multiple-objective transportation problem, *Naval Research Logistics Quarterly* 26(1): 123–139.

Kaufmann, A. and Gupta, M. M. (1988). *Fuzzy mathematical models in engineering and management science*. Elsevier Science Inc., Amsterdam.

Kratschmer, V. (2001). A unified approach to fuzzy random variables, *Fuzzy Sets and Systems* 123(1): 1–9.

Kwakernaak, H. (1978). Fuzzy random variables–I. Definitions and theorems, *Information Sciences* 15(1): 1–29.

Lai, Y. and Hwang, C. L. (1992). A new approach to some probabilistic linear programming problems, *Fuzzy Sets and Systems* 49: 121–123.

Lee, S. M. and Moore, L. J. (1973). Optimizing transportation problems with multiple objectives, *AIIE Transactions* 5(4): 333–338.

Liao, C.-N. (2009). Formulating the multi-segment goal programming, *Computers & Industrial Engineering* 56(1): 138–141.

Liu, B. and Iwamura, K. (1998). Chance constrained programming with fuzzy parameters, *Fuzzy Sets and Systems* 94(2): 227–237.

Luhandjula, M. (1983). Linear programming under randomness and fuzziness, *Fuzzy Sets and Systems* 10(1–3): 45–55.

Luhandjula, M. (2003). Mathematical programming in the presence of fuzzy quantities and random variables, *Journal of Fuzzy Mathematics* 11(1): 27–40.

Luhandjula, M. and Gupta, M. (1996). On fuzzy stochastic optimization, *Fuzzy Sets and Systems* 81(1): 47–55.

Nanda, S. and Kar, K. (1992). Convex fuzzy mappings, *Fuzzy Sets and Systems* 48(1): 129–132.

Nanda, S., Panda, G. and Dash, J. (2006). A new solution method for fuzzy chance constrained programming problem, *Fuzzy Optimization and Decision Making* 5(4): 355–370.

Nanda, S., Panda, G. and Dash, J. (2008). A new methodology for crisp equivalent of fuzzy chance constrained programming problem, *Fuzzy Optimization and Decision Making* 7(1): 59–74.

Panda, A. and Das, C. B. (2014). Multi-choice linear programming for matrix game, *Applied Mathematics and Computation* 237: 411–418.

Patro, K. K., Acharya, M. M., Biswal, M. and Acharya, S. (2015). Computation of a multi-choice goal programming problem, *Applied Mathematics and Computation* 271: 489–501.

Zadeh, L. (1968). Probability measures of fuzzy events, *J. Math. Anal. Appl* 23(2): 421–427.

Zarghami, M. (2010). Urban water management using fuzzy-probabilistic multi-objective programming with dynamic efficiency, *Water Resources Management* 24(15): 4491–4504.

Zimmermann, H. J. (1978). Fuzzy programming and linear programming with several objective functions, *Fuzzy Sets and Systems* 1(1): 45–55.

Chapter 4

A Multi-Objective Programming Approach for Disaster Management

Asipu Anusha, Sushree Swetapadma, Srikumar Acharya, M. M. Acharya, and S. K. Misra

4.1 INTRODUCTION

Man-made and natural disasters are inescapable in part of the global landscape. Several regions in India face significant disaster management challenges. Tropical cyclones are massive, rotating systems of storms that arise over warm ocean waters in tropical and subtropical regions (Emanuel, 2003). In the Atlantic and Northeast Pacific, these are recognized as hurricanes; in the Northwest Pacific, they are called typhoons; and in the Southern and Indian oceans, they are termed cyclones. Low pressure, strong winds, and plenty of rain are characteristics of these storms. The formation originates from a low-pressure system that rotates as the Earth rotates, creating an eye in the center, as warm, humid air rises from the ocean. Due to powerful winds, torrential rain, and storm surges that cause flooding, especially in coastal areas, they cause significant damage. Evacuation and preparedness are essential for preventing property and fatality losses. Meteorological agencies are carefully tracking these storms, which are most frequent in the Northern Hemisphere from June to November, for the purpose of providing early warnings and mitigating their effects on affected areas (Cheng, 2021)

Coastal regions around the world are particularly vulnerable to tropical cyclones, given their proximity to the ocean and their susceptibility to storm surges, heavy rainfall, and extreme winds. Given their geographical location, these regions are frequently already vulnerable to a variety of natural hazards, making them even more vulnerable (Balaguru, 2014). The northeastern trade winds blow across the tropical region that makes up the entire Indian peninsula. Additionally, the cyclones face the eastern coastal plain more frequently because they move from east to west. Odisha is particularly vulnerable to a variety of natural disasters. Situated on India's eastern coast, the state has frequently been hit by cyclones, floods, and even landslides, and extreme heat waves.

Odisha's geographical position along the Bay of Bengal makes it particularly vulnerable to cyclones, especially during the monsoon season. Among all the coastal states of India, Odisha is more prone to cyclones, with nearly

DOI: 10.1201/9781003642091-4

one-third of cyclones from the east coast affecting the state (Sarkhel, 2019). Of the total severe cyclonic storms of the Bay of Bengal, 15% affect Odisha, and districts like Balasore, Bhadrak, Jajpur, Cuttack, Puri, Ganjam, Kendrapara, Jagatsinghpur, Khordha, and Gajapati are more prone to cyclones. The state has experienced a number of significant cyclones, the most catastrophic of which was the super cyclone in 1999, which resulted in extensive damage and fatalities (Mishra, 2020). Additionally, floods have been a persistent threat in Odisha, especially in its low-lying regions, and droughts are common in its western districts. Predicting and effectively managing these events are difficult due to the complex topography and changing weather patterns.

Extensive preparedness and response plans developed in the wake of Odisha's experiences with frequent cyclones have provided the state with significant lessons learned from storms that have occurred over the years, especially the 1999 Super Cyclone, which immensely devastated the area and destroyed many thousand people (Das, 2024). Therefore, Odisha has invested in strengthening its early warning systems, which include warning local communities in real time about the impending storm. These systems forecast the trajectory and intensity of cyclones using current technology, including satellites and weather modeling. Due to this, local governments have been able to act quickly and warn people, and evacuate the vulnerable population well in advance. The construction of cyclone shelters has also massively reduced the number of casualties. These shelters, often built on higher ground, are designed to be resistant to strong winds and provide a place of safety for the communities impacted. Table 4.1 presents the data on cyclones that had severe impacts on the Odisha coast in the recent past.

Odisha's experience with tropical cyclones underlines the imperative of a multipronged and proactive approach in disaster management. The state's vulnerability to these storms, combined with its socioeconomic challenges, necessitates the incorporation of risk reduction, preparedness, and resilience building into all aspects of disaster response. By concentrating not only on immediate response but also on long-term solutions by implementing different disaster management programs, Odisha can better safeguard its communities, mitigate the loss of life and property, and strengthen its capacity to deal with the increasing frequency and severity of natural disasters in the future.

However, particularly in remote areas where the communication infrastructure is inadequate (cf. Figure 4.1), the challenges persist. Although evacuation protocols have been improved and maintained carefully, the logistical challenges of reaching distant populations can limit the effectiveness of these measures. Furthermore, the rapid expansion of urban areas, combined with poor land-use planning, has increased vulnerability in coastal cities, where informal settlements are frequently located in cyclone-prone areas. These areas are frequently overcrowded, lacking basic amenities, and underprepared to deal with the devastation of a major storm.

Table 4.1 Cyclone Occurrences in Odisha in the Recent Past.

Sl. No	*Name*	*Period*	*Category*	*Place of Landfall*	*Severely Affected Region*
11	Titli	October 10–12, 2018	Severe cyclonic storm	Palasa and South Odisha	South Odisha
2	Fani	May 3, 2019	Extremely severe cyclonic storm	Puri	East coastal district of Odisha; Puri, Kendrapara, Bhadrak, Khordha, Baleswar
3	Bulbul	November8–10, 2019	Very severe cyclonic storm	Sagar Island	Kendrapara, Bhadrak, Mayurbhanj, Baleswar
4	Amphan	May 20, 2020	Super cyclonic storm	Bakkhali, West Bengal	East coastal Odisha and northern Odisha
5	Yaas	May 25–26, 2021	Severe cyclonic storm	Near Baleswar	Bhadrak, Baleswar
6	Dana	October 24–25, 2024	Severe cyclonic storm	Habilikhati (Kendrapara)	Kendrapara, Bhadrak, Baleswar, Jagatsinghpur, Cuttack

Figure 4.1 Map highlighting coastal districts of Odisha impacted by tropical cyclones, illustrating vulnerability and response areas.

The prime purpose of the implementation of advanced technology in disaster management cannot be overstated. In Odisha, technological advancements have played a crucial role in enhancing the accuracy and timeliness of weather forecasts, leading to improved disaster preparedness (Das, 2024). The application of mobile apps, geographic information systems (GIS), and satellite imagery to track the movement of cyclones is significantly increased. Furthermore, in post-disaster damage assessments, drones and other unmanned aerial vehicles (UAVs) are also increasingly being utilized. Relief efforts can proceed more quickly as a result of authorities being able to access areas that might be challenging for conventional vehicles. Moreover, social media platforms and mobile phone alerts are examples of technological tools being utilized to spread information and increase public awareness. To make these technologies effective, community involvement is essential. The local communities are instructed and trained about evacuation plans, first aid, and early warning systems by state and district administrations, which can help people act quickly and reduce their risks. Communities often have a better grasp of their surroundings and weaknesses, so it is important for disaster management efforts to include local knowledge (Mohanty, 2022). Involving community members in disaster preparedness plans ensures that these plans are culturally appropriate and tailored to the specific needs of the population.

It is very important that disaster management in Odisha responds to the challenges posed by the increasing threats from tropical cyclones and other natural disasters. The traditional response model, which often relies on reactive measures, is no longer enough. A better approach to disaster management is needed due to the increasing frequency and severity of disasters. A multi-objective programming approach is essential for tackling both immediate and long-term challenges (Deb, 2016). This approach should involve a combination of risk reduction strategies, preparedness, response, recovery, and resilience building. It must integrate the efforts of local communities, governmental agencies, nongovernmental organizations, and international bodies (Cao, 2018). Whenever it addresses disaster management, the needs of individuals who are most affected, such as women, children, the elderly, and people with disabilities, must be prioritized foremost. Planning and response activities should involve these groups to guarantee that no one is left behind. During disasters, they frequently face additional risks. By empowering, educating, and fostering community resilience, Odisha can strengthen its capacity to endure and recover from cyclones and other natural disasters in the future.

A strategy that aims to accomplish multiple objectives simultaneously, each of which addresses various aspects of catastrophe risk, is known as a multi-objective approach to disaster management and mitigation (Krasko, 2017). In Odisha, disaster management generally employs a defensive strategy, emphasizing emergency response and recovery. However, a more proactive, all-encompassing structure is required due to the growing frequency and intensity of disasters. The multi-objective approach provides a way to handle the intricate and often conflicting priorities of resilience building, response, recovery, and risk mitigation (Panzo, 2019). This strategy seeks to maximize several areas of catastrophe management, ensuring that resources are used efficiently. It stresses the necessity of not only responding to immediate relief distribution (Noyan, 2012) but also addressing the underlying causes of vulnerability and developing adaptive capacity for the affected area. Furthermore, it emphasizes the importance of combining community-led activities with top-down policies and frameworks to boost local resilience and self-sufficiency (Grass, 2016). This chapter develops a mathematical model for post-disaster planning with human casualties, which can be considered as operational guidance for the proper use of emergency resources. For this purpose, a multi-objective linear programming model is provided to formulate the problem. The objective functions of the model aim to maximize proper relief distribution planning, maximize medical care facilities, and provide temporary accommodations in the affected area. A comprehensive computational analysis is performed for evaluation purposes. Also, a case study is made on the cyclone-affected coastal districts of Odisha, which illustrates the real-world application of the model.

4.2 STUDY REGION

Severe cyclonic storm Dana was a devastating tropical cyclone that impacted West Bengal and Odisha in India. Dana, the third and second severe cyclonic storm of the 2024 North Indian Ocean cyclone season, formed from a low-pressure area first observed by the Indian Meteorological Department (IMD) on October 20, 2024. On the basis of forecasts issued by the IMD and other international agencies such as the Regional Integrated Multi-Hazard Early Warning System (RIMES) and the Joint Typhoon Warning Center (JTWC), experts from the Odisha State Disaster Management Authority (OSDMA), a GIS platform, thoroughly tracked the system to determine the likelihood of its impact on Odisha.

As per IMD's bulletin, the cyclonic storm had moved north-westwards and crossed the north Odisha and West Bengal coast between Puri and Sagar Island close to Bhitarkanika and Dhamara (Odisha), during the midnight of October 24–25, 2024, as a severe cyclonic storm with wind speed 100–110 kmph and wind gust 120–125 kmph. It made landfall at Habalikhati Nature Camp (Bhitarkanika), Kendrapara, with a wind speed of 110–120 kmph and massive rainfall. The following image shows the track of Cyclone Dana, issued by the IMD (cf. Figure 4.2).

The average rainfall of the severely affected area is more than 150 mm. Bhadrak had recorded an average rainfall of 239.6 mm, and Kendrapara and Balasore recorded more than 200 mm rainfall (cf. Tables 4.2 and 4.3).

From the analysis, it is observed that Bhadrak, Balasore, and Kendrapara districts have been severely affected by the cyclonic storm with heavy rainfall and high wind speed. As these three districts were marked as red zones by IMD, their respective District Disaster Management Authorities (DDMAs) had remained in full preparedness to effectively deal with the situation.

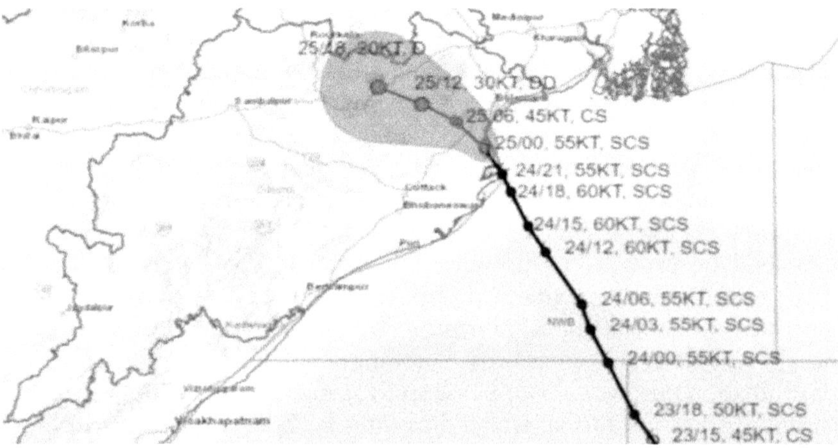

Figure 4.2 Map indicating the track of Cyclone Dana (forecasted by IMD).

Table 4.2 District-Wise Average Rainfall.

Sl. No.	*District*	*Average Rainfall*
1	Bhadrak	239.6 mm
2	Kendrapara	165.1 mm
3	Balasore	147.3 mm
4	Jajpur	134.8 mm

Table 4.3 Rainfall Status of Different Blocks Severely Affected.

Sl. No.	*District*	*Block*	*Total*
1	Balasore	Oupada	349.0 mm
2	Bhadrak	Chandbali	344.0 mm
3	Kendrapara	Rajkanika	380.0 mm
4	Bhadrak	Dhamanagar	257.0 mm

Table 4.4 People Evacuated.

Sl. No.	*District*	*No. of Shelter*	*No. of People*
1	Bhadrak	917	137,787
2	Balasore	1,213	212,940
3	Kendrapara	646	121,116

The Disaster Management department of Odisha with the district administration had arranged a multipurpose shelter house to accommodate evacuated people. Nearly 5 lakh people had been evacuated from the affected areas of Bhadrak, Balasore, and Kendrapara. Temporary shelter building (i.e., temporary accommodation had been identified and was ready for accommodation. Food, water supply, emergency healthcare, and other essentials like kerosene, candles, and polythene sheets were provided by the state government (cf. Table 4.4).

The damage caused by Cyclone Dana was mainly due to the gusting action of the wind and heavy rainfall. As per the report, these three districts need more attention for post-disaster management.

4.3 DISASTER RISK MITIGATION AND MANAGEMENT MODEL

In this study, multi-objective linear programming problem (Deb, 2016) has been used to analyze proper relief distribution in the affected area in the post-disaster scenario. Major disaster events are typically followed by multiple assessments carried out by the state government. These assessments are wide-ranging and require an immediate basis. For this reason, there is a growing consensus on the need for standardized and comprehensive assessment in a post-disaster scenario. Such assessment must be multi-sectoral, addressing proper food and water supply, emergency medical care, shelter, cooking wares, and social and community services in a balanced and comprehensive manner. The overarching purpose of this study is to provide proper relief distribution logistics for post-disaster management.

The multi-objective disaster management model is formulated as

$$\max f(x) = [f_1(x), f_2(x), f_3(x)],$$

$$\text{s.t. } x \in \mu,$$

$$\mu = \left\{ x \left| \begin{array}{l} a_i(x) \le 0, i = 1,2,3,\ldots n_1 \\ a_i(x) \le 0, i = 1,2,3,\ldots\ n_2 \\ x = [x_1, x_2, x_3 \ldots, x_n]^T \end{array} \right. \right\}.$$

Here, x_i = decision variable,
x = solution of the model, and
$f(x)$ = linear objective function.

4.4 FORMULATION OF THE MODEL

To enhance the fund supply and appropriate relief management technique, the optimization model is represented as

$$\max f(x) = [f_1(x), f_2(x), f_3(x)].$$

The proposed multi-objective problem usually comprises two main steps: generalization of efficient solutions and identification of the best compromise solution, which should be an efficient solution (Giagkiozis, 2015). Furthermore, a reasonable fund supply has been determined to improve basic needs and healthcare facilities for the affected people. The model suggested the proper distribution of funds, which can help the district administration manage post-disaster troubles.

The model works on the constraints represented as

$$\begin{array}{l} a_i(x) \le 0, i = 1,2,3,\ldots n_1 \\ a_i(x) \le 0, i = 1,2,3,\ldots\ n_2 \\ x = [x_1, x_2, x_3 \ldots, x_n]^T \end{array}.$$

The people evacuated from severely affected regions are taken into consideration, and the facilities provided by the government, such as food supply, medical care, other essentials, and allocation of temporary accommodation, are considered as decision variables. Therefore, we can write

$$Max : [f_1 = 1.3x_1 + 0.8 \quad x_2 + 0.6 \quad x_3 + 0.2 \quad x_{4,} f_2 = 2.1 \quad x_1 + 1.01 \quad x_2 + 0.7 \quad x_3 + 0.4 \quad x_{4,} f_3 = 1.2\, x_1 + 0.65\, x_2 + 0.42\, x_3 + 0.2\, x_4].$$

Subject to

$$x_1 + x_2 + x_3 + x_4 < 700;$$

$$x_1 + x_2 \leq 400;$$

$$x_2 + x_3 \leq 100;$$

$$x_1 + x_3 \leq 500.$$

The basic idea of the proposed model is first to design an algorithm to solve for the optimized value of the height parameter corresponding to the time-efficient objectives (Caramia, 2020). The purpose of the computational experiment is to compare the funding supply given by the state government and the relief logistics provided by the model. First, the model aims to make a framework for a relief distribution in an emergency situation within a limited fund. According to the data collected, the model can obtain a total effective response under a post-disaster scenario.

4.5 RESULTS AND DISCUSSION

An efficient multi-objective optimization problem has been formulated to achieve the prioritized multiple objectives as much as possible. The well-known weighting method has been used to solve the model. The addressed problem not only improves relief fund distribution by considering different factors but also analyzes the risk mitigation significantly (Wang, 2016). The result is verified by Python programming for better analysis. Table 4.5 shows the fund allocation for relief supplies in the affected area.

Table 4.5 Fund Allocation for Relief Supply in the Affected Area.

Sl. No.	*Name of the District*	*Total Relief Fund Allocation (in Lakhs)*
1	Bhadrak	635
2	Balasore	961
3	Kendrapara	568

After the disaster, the state and district administrations focused on food and drinking water supply to the affected people. Odisha has a robust disaster management system with pre-stocked relief material at cyclone shelters across the state. Beyond immediate food supply, the government should focus on the medical and healthcare of people. Special medical attention should be given to pregnant women, children, the elderly, and people with disabilities during the relief operation. In general, government and various agencies neglected medical care post-disaster management. Therefore, after the cyclone, the affected people suffered from various types of health issues like fever, common cold, and diarrhea. Also, in the shelter house, the evacuated people suffered from many health problems. The proposed model suggests that there should be an emergency medical care fund to serve people as soon as possible. Emergency relief supplies also include other essentials like kitchen wares, clothing, plastic sheeting, and lanterns. These items were supplied and distributed to the villagers. Keeping in view of its necessities, the special relief commission focused on the preparedness of essential items in the cyclone shelter. The model also emphasizes that the fund supply for essential items should be maximized to maintain proper care of the affected people.

As the number of evacuated people is higher, the cyclone shelter is not sufficient to accommodate them. Most of the blocks of Bhadrak, Balasore, and Kendrapara lie on the sea shore, which are severely affected by cyclones each year. So, to accommodate the large number of people, temporary accommodation should be installed. On the other hand, transportation of affected people also poses a challenge for the administration. Therefore, the mathematical model suggests that the number of temporary accommodations should increase and the fund supply for the management of accommodating buildings should be higher. As a preventive measure, nearby schools, community centers, and Anganwadi centers are used as temporary accommodation. In order to streamline and improve the quality of management, upkeep, and maintenance of temporary accommodation, the State Disaster Management Department should provide funds.

According to the above results, the advantages of the application of the multi-objective optimization method are verified for emergency decision-making. After taking human factors into consideration, fund supply parameters were introduced for the accuracy of the model. The problems brought by increasing the accuracy include the solving time and effectiveness of the algorithm. However, the improvement of precision is helpful to improve the quality of the result.

4.6 CONCLUSIONS

The operational studies theory and method have been employed to solve a multi-objective optimization problem and generate a scientific model for relief distribution in post-disaster situations. The prime objective of the study is to establish an economic cost–benefit relief distribution logistic, which involves relevant fund supply and management of the economy in an emergency situation. The model refers to the fund supply management that plays a vital part in the relief approach (Mishra, 2020). In order to enhance the effectiveness of model solving, the algorithm's primary parameters were also examined. The computation results of a case-based study were implemented to confirm the accuracy and the advantages of the suggested model.

REFERENCES

Balaguru, K. T. (2014). Increase in the intensity of postmonsoon Bay of Bengal tropical cyclones. *Geophysical Research Letters*, 41(10), 3594–3601.

Cao, C. L. (2018). A novel multi-objective programming model of relief distribution for sustainable disaster supply chain in large-scale natural disasters. *Journal of Cleaner Production*, 174, 1422–1435.

Caramia, M. D. (2020). Multi-objective optimization. In *Multi-objective management in freight logistics. Increasing capacity, service level, sustainability, and safety with optimization algorithms* (pp. 21–51). Springer International Publishing, Cham.

Cheng, J. F. (2021). Modeling the equitable and effective distribution problem in humanitarian relief logistics by robust goal programming. *Computers & Industrial Engineering*, 155, 107–183.

Das, T. S. (2024). Assessing tropical cyclone risk for improving mitigation strategies in Coastal Odisha, India. *Environmental Science and Pollution Research*, 1–21.

Deb, K. S. (2016). Multi-objective optimization. In *Decision sciences* (pp. 161–200). CRC Press, Boca Raton.

Emanuel, K. (2003). Tropical cyclones. *Annual Review of Earth and Planetary Sciences*, 31(1), 75–104.

Giagkiozis, I. (2015). Methods for multi-objective optimization: An analysis. *Information Sciences*, 293, 338–350.

Grass, E. (2016). Two-stage stochastic programming in disaster management: A literature survey. *Surveys in Operations Research and Management Science*, 21(2), 85–100.

Krasko, V. (2017). Two-stage stochastic mixed-integer nonlinear programming model for post-wildfire debris flow hazard management: Mitigation and emergency evacuation. *European Journal of Operational Research*, 263(1), 265–282.

Mishra, M. P. (2020). Analyzing the trend of tropical cyclone activity along Odisha Coast. *Progress in Disaster Science*, 100092.

Mohanty, A. D. (2022). Policy and governance strategies for effective cyclone risk management in Odisha, India: A journey from 1999 super cyclone. In *Cyclonic Disasters and Resilience: An Empirical Study on South Asian Coastal Regions* (pp. 155–184). Springer Nature Singapore, Singapore.

Noyan, N. (2012). Risk-averse two-stage stochastic programming with an application to disaster management. *Computers & Operations Research*, 39(3), 541–559.

Panzo, A. S. (2019). Disaster relief model in laguna using integer linear programming, in mobile and wireless technology 2018. In *International Conference on Mobile and Wireless Technology (ICMWT 2018)* (pp. 463–470). Springer Singapore, Singapore.

Sarkhel, P. B. (2019). A review of cyclone and its impact on the coastal belts of Odisha. Sarkhel, P., Biswas, D., & Swain, S. S. (2019). *A Review of International Journal of Engineering Research \& Technology*, 8(5), 759–762.

Wang, R. Z. (2016). Localized weighted sum method for many-objective optimization. *IEEE Transactions on Evolutionary Computation*, 22(1), 3–18.

Chapter 5

Trade Credit Based Inventory Model Using Credit-Linked Demand

Alok Kumar

5.1 INTRODUCTION AND LITERATURE REVIEW

Trade credit period is an important marketing tool to attract customers who are either suppliers or retailers. It becomes more crucial when a new product is launched into the market. This chapter develops an inventory model where the demand of the product is based upon the innovation diffusion criterion, and the credit period is given by the retailer to the customers. There are several inventory models that have incorporated the effect of the trade credit period to influence the demand. A few of them are discussed here as follows. Chang et al. (2003) developed a time-dependent demand function economic order quantity model where the items are deteriorating and the supplier provides a credit period to the retailer. The credit period given by the supplier to the retailer and the retailer's promotional effort strategy for the inventory replenishment problem, where demand is a function of price and time, were discussed by Tsao and Sheen (2008). The firm's credit policy was incorporated using a discounted cash flow approach to obtain the optimum order quantity by Chung and Liao (2009). Thangam and Uthayakumar (2009) developed an inventory model under permissible delay in payments, where demand is a function of unit selling price and the credit period given by the retailer to the customer. In a similar fashion, an integrated vendor–buyer inventory system was developed by Chang et al. (2009), where the demand rate is assumed to be a decreasing function of the retail price and the trade credit is linked to the order quantity. An optimum order quantity model where demand is time dependent and linearly increasing, and the two levels of trade credit, i.e., credit period given by the supplier to the retailer, as well as the retailer to the customer, were developed by Teng et al. (2013).

Jaggi et al. (2013) developed an inventory model where the items are of imperfect quality and the screening rate is greater than the demand rate under the permissible delay in payments, and also the shortages are allowed and fully backlogged. An economic order quantity model where the product is of a deteriorating nature and the demand rate is a function of time, and

DOI: 10.1201/9781003642091-5

incorporates the effect of the length of the credit period, was discussed by Yang et al. (2015). A two-level partial trade credit-based economic order quantity model was developed by Mondal et al. (2015), where the demand is dependent upon the inventory level and the credit period. A three-level partial trade credit supply chain inventory model where there is a single manufacturer, a single distributor, and a single retailer was developed by Singh et al. (2015). Ouyang et al. (2015) developed an inventory model for permissible delay in the payment period with capacity constraint. An inventory model with partial and full trade credit policy, where suppliers offer full trade credit to the retailers and the retailers offer only partial trade credit to their customers, was developed by Sarkar et al. (2015). There are a few inventory models that are based upon the innovation diffusion-dependent demand function and have incorporated the effects of permissible delay in the payment period. These models are discussed as follows. To discuss the inventory lot-size policies, Chern et al. (2001) developed an inventory model for new durable products where demand is based upon the Bass (1969) model. Kumar and Aggarwal (2012) developed an inventory model that is based upon an innovation diffusion process under permissible delay in payments, where the supplier offers a credit period to the retailers and there is no credit period given by the retailer to their customers. In this model, the demand function is associated with two kinds of influences: internal influences and external influences. Shah (2015) developed a manufacturer–retailer inventory model where the demand is price-sensitive and credit-linked under the environment of two levels of trade credit financing and also under the effect of a profit-sharing contract. An inventory model with permissible delay in payments where the demand is dependent upon the dynamic pricing-advertising condition was discussed by Kumar and Chanda (2017a)—an economic order quantity model where demand is bifurcated into two parts, one that is based upon the external influence, i.e., the innovation effect, and the other that takes into account the effects of external and internal influences.

In this model, the case of permissible delay in payments has also been taken, where the supplier offers a credit period to the retailers. Jaggi et al. (2019) discussed a policy for the inventory decisions for the deteriorating items where the demand is stock-dependent using Stackelberg and Nash equilibrium solutions under the environment of credit policy. Sundararajan et al. (2021) developed an inventory model for price- and time-dependent demand using inflation and delay in payment, where the case of backlogging has also been considered. Nagpal et al. (2022) developed an inventory model where the generation of technology products has been emphasized by incorporating the trade credit effects and the imprecise procurement costs. Kumar (2022) considered a model where the dynamic effect of advertising and pricing has been incorporated in the demand model, and the parameters associated with these advertising and pricing policies have been taken as fuzzy variables. Bhattacharjee et al. (2023) discussed a single-manufacturer and a single-retailer supply chain model under the two levels of trade credit policy for a payment time-dependent discount factor, and the market demand is linearly dependent on the selling price. Tiwari et al. (2024) proposed an economic production quantity model for the price-dependent demand rate and the partial trade credit policy with inflation under the cloudy fuzzy environment. Wang et al. (2025) discussed a multi-item inventory model under a cash discount and trade credit policy that includes both a standard payment term and an incentive for early payment in the form of a cash discount. This chapter discusses an economic order quantity model where the demand for the product is influenced by the innovation diffusion criterion, and it is also credit-linked. To understand the nature of the model, a numerical example is well explained in Section 5.4. The chapter has been organized into several sections as given below. The model, which is based on certain assumptions and the notations used to develop the model, is given in Section 5.2. The mathematical model is formulated in Section 5.3. The optimum solutions with respect to different parameters are obtained in Section 5.4. Observations regarding the behavior of the model are given in Section 5.5. In Section 5.6, the managerial implications and the conclusion are given.

5.2 ASSUMPTIONS AND NOTATIONS

The inventory model is always based upon certain assumptions, and notations are used to develop the mathematical model that are detailed as follows. The supplier offers a credit period (M) to the retailer, and the retailer also offers a credit period (S) to their customers. Demand rate is time-dependent and is based upon the innovation diffusion criterion. The innovation diffusion criterion is based upon the external influence, which is the innovation effect. The innovation effect (z) and the potential market size (N) are constant throughout the cycle time. There is an instantaneous replenishment rate with negligible lead time, and shortages are not allowed to occur. Ordering cost per order (A), cost per unit item (C), inventory carrying charge (I), profit per unit item (P), interest charged per dollar per unit time (I_c), and interest earned per dollar per unit time by the retailer (I_e) are known in advance. Here, \$ is the unit for price or cost. In addition, the following notations have been used.

$I(t)$: Inventory level at any time t
T: Replenishment cycle length
Q: Inventory at the beginning of the period
$K(T)$: Total cost per unit time
$\lambda(t)$: Demand function
$N(t)$: Cumulative number of adopters at time t.

5.3 FORMULATION OF THE MATHEMATICAL MODEL

The demand function for the model is described as follows:

$$\lambda(t) = \lambda_1(t) = Nze^{-zt} + \alpha(S-t),\ 0 \le t \le S, \tag{5.1}$$

$$\lambda_2(t) = Nze^{-zt} \qquad S \le t \le T. \tag{5.2}$$

The lot size (Q) at the beginning of the cycle time (T) is

$$Q = \int_0^T \lambda(t)dt = Q_1 + Q_2 = \int_0^S \lambda_1(t)dt + \int_S^T \lambda_2(t)dt, \tag{5.3}$$

$$\text{where } Q_1 = \int_0^S \lambda_1(t)dt \text{ and } Q_2 = \int_S^T \lambda_2(t)dt, \tag{5.4}$$

$$\Rightarrow Q = N\left(1-e^{-zS}\right) + \frac{\alpha S^2}{2} + N\left(e^{-zS} - e^{-zT}\right). \tag{5.5}$$

Inventory level at any time t during the cycle time T is

$$I(t) = I_1(t) = Q - \int_0^t \lambda_1(t)dt, 0 \le t \le S, \tag{5.6}$$

$$I_2(t) = Q_2 - \int_S^t \lambda_2(t)dt, S \le t \le T, \tag{5.7}$$

$$\Rightarrow I_1(t) = N\left(1-e^{-zS}\right) + \frac{\alpha S^2}{2} + N\left(e^{-zS} - e^{-zT}\right) + N\left(e^{-zt} - 1\right) - \alpha\left(St - \frac{t^2}{2}\right), \tag{5.8}$$

$$\text{and } I_2(t) = N\left(e^{-zt} - e^{-zT}\right). \tag{5.9}$$

The cost components are given as follows:

$$\text{Ordering cost per unit time } (OC) = \frac{A}{T}. \tag{5.10}$$

Purchasing cost per unit time (PC) is

$$PC = \frac{CQ}{T}, \tag{5.11}$$

$$\Rightarrow PC = \left(\frac{C}{T}\right)N\left(1-e^{-zS}\right)+\frac{C\alpha S^2}{2T}+\left(\frac{C}{T}\right)N\left(e^{-zS}-e^{-zT}\right). \tag{5.12}$$

Inventory carrying cost per unit time (ICC) is

$$ICC = \left(\frac{IC}{T}\right)\left(\int_0^S I_1(t)\,dt+\int_S^T I_2(t)\,dt\right), \tag{5.13}$$

$$\begin{aligned}\Rightarrow ICC = {} & \frac{ICN}{T}(1-e^{-zS})\left(S+\frac{1}{z}\right)-\frac{ICNS}{T}+\frac{IC\alpha S^3}{6T}+ \\ & \frac{ICNS}{T}(e^{-zS}-e^{-zT})+\frac{ICN}{zT}\left(e^{-zS}-e^{-zT}\right)- \\ & \frac{ICN}{T}(T-S)e^{-zT}.\end{aligned} \tag{5.14}$$

Case (i) $M \le T$

In this case, the retailer earns the interest at the rate of I_e during the time periods $(0, M)$ and (S, M), and the interest is charged to the retailer at the rate of I_c during the period (M, T).

Interest earned (IE_1) by the retailer is

$$IE_1 = \frac{PI_e}{T}\left(\int_0^M tNze^{-zt}\,dt+\int_S^M \frac{\alpha S^2}{2}\,dt\right), \tag{5.15}$$

$$\Rightarrow IE_1 = \frac{PI_e\alpha S^2}{2T}(M-S)+\frac{PI_eN}{zT}\left(1-e^{-zM}\right)-\frac{PI_eNMe^{-zM}}{T}. \tag{5.16}$$

Interest charged (IC) is

$$IC = \frac{CI_c}{T}\int_M^T I_2(t)\,dt, \tag{5.17}$$

$$\Rightarrow IC = \frac{CI_cN\left(e^{-zM}-e^{-zT}\right)}{zT}-\frac{CI_cNe^{-zT}(T-M)}{T}. \tag{5.18}$$

The total cost per unit time $K_1(T)$ is

$$K_1(T) = OC+PC+ICC-IE_1+IC, \tag{5.19}$$

$$\Rightarrow K_1(T)=\frac{A}{T}+\left(\frac{C}{T}\right)N\left(1-e^{-zS}\right)+\frac{C\alpha S^2}{2T}+$$
$$\left(\frac{C}{T}\right)N\left(e^{-zS}-e^{-zT}\right)+\frac{ICN}{T}\left(1-e^{-zS}\right)\left(S+\frac{1}{z}\right)-$$
$$\frac{ICNS}{T}+\frac{IC\alpha S^3}{6T}+\frac{ICNS}{T}\left(e^{-zS}-e^{-zT}\right)+ \tag{5.20}$$
$$\frac{ICN}{zT}\left(e^{-zS}-e^{-zT}\right)-\frac{ICN}{T}(T-S)e^{-zT}-\frac{PI_e\alpha S^2}{2T}(M-S)-$$
$$\frac{PI_eN}{zT}\left(1-e^{-zM}\right)+\frac{PI_eNMe^{-zM}}{T}+\frac{CI_cN\left(e^{-zM}-e^{-zT}\right)}{zT}-$$
$$\frac{CI_cNe^{-zT}(T-M)}{T}.$$

Case (ii) $T \leq M$

In this case, interest is earned by the retailer throughout the cycle time T, and no interest is charged to the retailer.

Interest earned (IE_2) by the retailer is

$$IE_2=\frac{PI_e}{T}\left(\int_0^T tNze^{-zt}dt+\int_S^T\frac{\alpha S^2}{2}dt+\int_T^M Qdt\right), \tag{5.21}$$

$$\Rightarrow IE_2=\frac{PI_eN}{zT}\left(1-e^{-zT}\right)-PI_eNe^{-zT}+\frac{PI_e\alpha S^2}{2T}(T-S)+$$
$$\frac{PI_eQ}{T}(M-T). \tag{5.22}$$

The total cost per unit time $K_2(T)$ is

$$K_2(T)=OC+PC+ICC-IE_2, \tag{5.23}$$

$$\Rightarrow K_2(T)=\frac{A}{T}+\left(\frac{C}{T}\right)N\left(1-e^{-zS}\right)+\frac{C\alpha S^2}{2T}+$$
$$\left(\frac{C}{T}\right)N\left(e^{-zS}-e^{-zT}\right)+\frac{ICN}{T}\left(1-e^{-zS}\right)\left(S+\frac{1}{z}\right)-\frac{ICNS}{T}+.$$
$$\frac{IC\alpha S^3}{6T}+\frac{ICNS}{T}\left(e^{-zS}-e^{-zT}\right)+\frac{ICN}{zT}\left(e^{-zS}-e^{-zT}\right)-$$

$$\frac{ICN}{T}(T-S)e^{-zT}-\frac{PI_eN}{zT}\left(1-e^{-zT}\right)+ \\ PI_eNe^{-zT}-\frac{PI_e\alpha S^2}{2T}(T-S)-\frac{PI_eQ}{T}(M-T) \tag{5.24}$$

The objective of the above two cases is to obtain the optimum value of T, which minimizes the total cost per unit time $K(T)$, where $K(T)$ is defined as follows:

$$K(T)=K_1(T) \text{ if } M\leq T, \tag{5.25}$$

$$K_2(T) \text{ if } M\leq T. \tag{5.26}$$

The optimum solutions will be obtained on the basis of Equations (5.20) and (5.24), which will be based on the conditions mentioned in Equations (5.25) and (5.26). It can be observed that the cost functions depicted in Equations (5.20) and (5.24) are very complex; therefore, it becomes difficult to obtain the convexity of the cost functions for the global optimal solution using an analytical method. The optimum solutions for different situations with respect to different parameters are explained below Section 5.4.

5.4 NUMERICAL EXAMPLE

In this section, sensitivity analysis with respect to different parameters has been well explained to understand the nature and behavior of the model. The sensitivity analysis results are given in Tables 5.1–5.3.

Table 5.1 Sensitivity Analysis with Respect to the Innovation Effect (z).

$A=1100$, $C=500$, $N=15000$, $P=1500$, $I=0.3$, $\alpha=1000$, $I_e=0.06$, $I_c=0.1$, $S=\frac{60}{365}$, $M=\frac{90}{365}$

z	T^*	$K(T^*)$	$Q(T^*)$
0.02	0.800	174,885	251.67
0.03	0.682	255,966	317.52
0.04	0.618	336,086	380.03
0.05	0.580	415,521	442.28
0.06	0.556	494,405	506.25

Table 5.2 Sensitivity Analysis with Respect to the Credit Period Offered by the Supplier to the Retailer (M).

$A = 1100,\ C = 500,\ N = 15000,\ P = 1500,\ I = 0.3,\ \alpha = 1000,\ I_e = 0.06,\ I_c = 0.1,$

$S = \frac{30}{365},\ z = 0.01$

M	T^*	$K(T^*)$	$Q(T^*)$
45/365	0.636	84,931	98.56
60/365	0.642	85,313	99.37
90/365	0.655	86,122	101.45
100/365	0.661	86,404	102.28
120/365	0.673	86,986	104.12

Table 5.3 Sensitivity Analysis with Respect to the Credit Period Offered by the Retailer to the Customer (S).

$A = 1100,\ C = 500,\ N = 15000,\ P = 1500,\ I = 0.3,\ \alpha = 1000,\ I_e = 0.06,\ I_c = 0.1,$

$M = \frac{120}{365},\ z = 0.01$

S	T^*	$K(T^*)$	$Q(T^*)$
30/365	0.673	86,986	104.12
45/365	0.877	89,835	138.63
60/365	1.100	92,948	177.72
90/365	1.570	99,483	264.49
100/365	1.730	101,699	295.41

In Table 5.1 shows that as the coefficient of innovation, i.e., innovation effect, increases, the optimum cycle length decreases, while the optimum lot size and the optimum cost increase. This means the inventory manager must procure more inventory for less time period to maintain the optimum cost when the innovation effect increases.

Table 5.2 shows that as the credit period offered by the supplier to the retailer increases, the optimum cycle length, the optimum lot size, and the optimum cost increase. This forces the inventory manager to make the optimum decisions based on the above trends.

Table 5.3 shows that as the credit period offered by the retailer to the customer increases, the optimum cycle length, the optimum lot size, and the optimum cost increase. This forces the inventory manager to make the optimum decisions based on the above trends.

5.5 OBSERVATIONS

In this section, several observations have been made on the basis of the sensitivity analysis results depicted in the aforementioned tables. The sensitive behavior of the parameters, such as the coefficient of innovation, credit period offered by the supplier to the retailer, and credit period offered by the retailer to the supplier, which play a significant role in making optimum decisions regarding the optimum inventory policies, are presented in the aforementioned tables. The behavior of the optimum cycle length, the optimum lot size, and the optimum cost can be well experienced when the sensitive nature of the parameters changes with respect to the environment under consideration.

5.6 CONCLUSION AND MANAGERIAL IMPLICATIONS

Credit-linked demand is one of the most important tools to attract customers when a new product is launched into the market. The nature and behavior of the model have been well understood using the sensitivity analysis with respect to several parameters, as discussed in Section 5.4. The sensitivity analysis also helps in making the optimum decisions in different situations. Table 5.1 shows that as the coefficient of innovation, i.e., the effect of external influence, increases by keeping other parameters constant, the optimum cycle length to procure the inventory decreases while the total optimum cost and the total optimum lot size increase. But a similar result is not seen in the case of the credit period offered by the supplier to the retailer and the retailer to the customer, as can be observed in Tables 5.2 and 5.3. That is, when the credit period for the retailer and the customer increases, the optimum cycle time, the optimum cost, and the optimum lot size all increase. Therefore, while taking optimum decisions regarding the environment as discussed above, this model can be effective for the inventory manager. This chapter can be extended for quantity discounts, a multi-item inventory system, and stochastic demand.

5.7 ACKNOWLEDGMENT

The infrastructural support provided by the FORE School of Management, New Delhi, in completing this chapter is gratefully acknowledged.

REFERENCES

Bass, F. M. (1969). A new-product growth model for consumer durables. *Management Science*, *15*, 215–227.

Bhattacharjee, R., Maiti, T., & Giri, B. C. (2023). Manufacturer–retailer supply chain model with payment time-dependent discount factor under two-levels of trade credit. *International Journal of Systems Science: Operations & Logistics*, *10*(1), 2005842.

Chang, C. T., Ouyang, L. Y., & Teng, J. T. (2003). An EOQ model for deteriorating items under supplier credits linked to ordering quantity. *Applied Mathematical Modelling*, *27*(12), 983–996.

Chang, H. C., Ho, C. H., Ouyang, L. Y., & Su, C. H. (2009) The optimal pricing and ordering policy for an integrated inventory model when trade credit linked to order quantity. *Applied Mathematical Modelling*, *33*(7), 2978–2991.

Chern, M. S., Teng, J. T., & Yang, H. L. (2001). Inventory lot-size policies for the bass diffusion demand models of new durable products. *Journal of the Chinese Institute of Engineers*, *24*, 237–244.

Chung, K.-J., & Liao, J.-J. (2009) The optimal ordering policy of the EOQ model under trade credit depending on the ordering quantity from the DCF approach. *European Journal of Operational Research*, *196*(2), 563–568.

Jaggi, C. K., Goel, S. K., & Mittal, M. (2013). Credit financing in economic ordering policies for defective items with allowable shortages. *Applied Mathematics and Computation*, *219*(10), 5268–5282.

Jaggi, C. K., Gupta, M., Kausar, A., & Tiwari, S. (2019). Inventory and credit decisions for deteriorating items with displayed stock dependent demand in two-echelon supply chain using Stackelberg and Nash equilibrium solution. *Annals of Operations Research*, *274*, 309–329.

Kumar, A. (2022). Dynamic pricing dependent demand under fuzzy criterion for an economic order quantity model. *Abhigyan*, *40*(2), 30–41.

Kumar, A., & Aggarwal, K. K. (2012). Optimal replenishment policy for an inventory model with demand follows innovation diffusion process under permissible delay in payments. *International Journal of Information and Management Sciences*, 177–197.

Kumar, A., & Chanda, U. (2017a). Economic order quantity under permissible delay in payments for new products in dynamic pricing-advertising condition. *International Journal of Business Innovation and Research*, *13*(2), 203–221.

Mondal, M., Maiti, M., & Maiti, M. (2015). EOQ model with variable demand and partial trade credit in fuzzy rough environment. *Asian Journal of Mathematics and Computer Research*, *9*(3), 273–301.

Nagpal, G., Chanda, U., Kumar, A., & Jasti, N. (2022). Inventory Modelling for technology generation products under uncertain trade credit terms and imprecise procurement costs. *Operations and Supply Chain Management: An International Journal*, *15*(3), 328–344.

Ouyang, L.-Y., Ho, C.-H., Su, C.-H., & Yang, C.-T. (2015) An integrated inventory model with capacity constraint and order-size dependent trade credit. *Computers & Industrial Engineering*, *84*, 133–143.

Sarkar, B., Saren, S., & Cárdenas-Barrón, L.E. (2015) An inventory model with trade-credit policy and variable deterioration for fixed lifetime products. *Annals of Operations Research*, *229*(1), 677–702.

Shah, N. H. (2015). Manufacturer-retailer inventory model for deteriorating items with price-sensitive credit-linked demand under two-level trade credit financing and profit sharing contract. *Cogent Engineering*, *2*(1), 1012989.

Singh, N., Vaish, B., & Singh, S. R. (2015) Three level supply chain model with variable demand rate under partial trade credit policy. *International Journal of Services and Operations Management*, *21*(4), 479–503.

Sundararajan, R., Vaithyasubramanian, S., & Nagarajan, A. (2021). Impact of delay in payment, shortage and inflation on an EOQ model with bivariate demand. *Journal of Management Analytics*, *8*(2), 267–294.

Teng, J. T., Yang, H. L., & Chern, M. S. (2013). An inventory model for increasing demand under two levels of trade credit linked to order quantity. *Applied Mathematical Modelling*, *37*(14–15), 7624–7632.

Thangam, A., & Uthayakumar, R. (2009). Two-echelon trade credit financing for perishable items in a supply chain when demand depends on both selling price and credit period. *Computers & Industrial Engineering*, *57*(3), 773–786.

Tiwari, S., Shah, K., & Bhimani, K. (2024). EPQ model with the effect of inflation and reliability for partial trade credit under fuzzy and cloudy fuzzy environment. *Journal of Management Analytics*, *11*(1), 110–134.

Tsao, Y. C., & Sheen, G. J. (2008). Dynamic pricing, promotion and replenishment policies for a deteriorating item under permissible delay in payments. *Computers & Operations Research*, *35*(11), 3562–3580.

Wang, L., Pi, Y., Zhang, Z., & Wang, S. (2025). Optimising the capacitated joint replenishment and delivery policies under cash discount and trade credit. *International Journal of Systems Science: Operations & Logistics*, *12*(1), 2466819.

Yang, C. T., Dye, C. Y., & Ding, J. F. (2015). Optimal dynamic trade credit and preservation technology allocation for a deteriorating inventory model. *Computers & Industrial Engineering*, *87*, 356–369.

Chapter 6

Analysis of Investment Pattern of DIIs and Its Impact on Nifty Returns during Pre- and Post-Recession Periods

P. Ajith Kumar and Topunuru Kaladhar

6.1 INTRODUCTION

Domestic institutional investors (DIIs) are investors who undertake investments in various securities and financial assets in a recognized stock exchange of the country where it is located. In simple terms, DIIs are investors who pool funds to trade in securities and assets in their own country. DIIs play a vital role in mobilizing savings and channelizing them into capital market. Increase in domestic investments has been one of the biggest contributions to the Indian economic growth, and both the public and private sectors have enabled and sustained these investments. In 2024, DIIs sharply increased their share in the Indian capital market, reached an all-time high of 16.90% by December 31, 2024, from 16.46% on September 30, 2024, as they invested nearly Rs 1.86 lakh crore (Internet Source).

6.1.1 Investors Driving the Domestic Investment in the Country

1. Public Sector Enterprises
2. Banks
3. Insurance companies and other financial institutions

6.2 REVIEW OF LITERATURE

Many institutions and individuals have conducted several studies on the various aspects of the securities market in the past. However, not many studies were found focusing on the investment pattern of DIIs and their impact on Nifty returns. Hence, a modest attempt is made here to review some of the studies relevant to the subject in order to obtain deeper insights into and understanding of the research area.

DOI: 10.1201/9781003642091-6

- **Vikram et al.** (2025)[1] found that foreign institutional investors (FIIs) and DIIs have a significant impact on Nifty 50 and Nifty Midcap 50. The study also suggested that it will helpful to institutional investors and retail investors to identify and compare different induced volatility.
- **Sreenivasula Sunkara et al.** (2025)[2] found that the performance of a nation's security markets is influenced by a complex set of factors including economic conditions, political stability, legal governance, government spending, foreign and domestic investments, trade policies, sectoral performance, and currency fluctuations. Among these, FII activity and DII activity are identified as critical drivers of equity market movements in India.
- **Swapna et al.** (2024)[3] examined and found that there is a positive correlation among FIIs, DIIs, and Nifty returns and also revealed that FIIs and DIIs show a significant impact on retail investors and their investing psychology. Furthermore, the study concluded that FIIs and DIIs have an inverse relationship with each other.
- **Rathod et al.** (2024)[4] found that FIIs are often linked to short-term volatility due to their rapid capital movements, while DIIs are considered stabilizing forces in the market. The study showed moderate to strong correlations between these investment flows and index returns, with combined effects explaining a substantial share of market variation. Empirical evidence frequently supports the use of regression models, provided normality and multicollinearity assumptions are addressed. In line with these findings, the present study confirms that FIIs and DIIs together explain 67.5% of Nifty 50 variation. This reinforces the established view that institutional flows are key predictors of market trends in emerging economies like India.
- **Gandhi and Ayre** (2024)[5] investigated impact of FIIs and DIIs on the Indian stock market and found that FIIs and DIIs have an impact on BSE Sensex variations.
- **Mathew and John** (2019)[6] revealed that market fluctuations are often driven by investor sentiment, particularly the withdrawal of foreign investors seeking better returns elsewhere. Research highlights the untapped potential of India's mutual fund industry, which remains underutilized despite awareness initiatives. Limited success in investor education suggests a need for stronger collaborative efforts between market participants and the government. Previous findings consistently indicate that enhanced DII engagement can stabilize the market. These insights underline the importance of policy measures to foster institutional investment for sustainable market growth.
- **S. A. Atif Salar and Mohd Shamim** (2017)[7] showed that DIIs have an influence on the movement of Nifty and suggested that a strong positive relation exists between stock market movement and DII. In their

study, they described that the Indian stock market has an Arch and Garch effect, which means that volatility in the Indian stock market is caused by the flow of the previous day's information and volatility. They also found that DIIs also contribute to an increase in volatility in the Indian stock market.

- **Tanu Aggarwal and Priya Solomon** (2017)[8] found out trends in DIIs and FIIs and how the BSE Sensex was influenced by the foreign direct investment (FDI) and FIIs during the study period from 2006–2007 to 2015–2016. To analyze the data during the study period, they used statistical tools such as trend analysis, correlation, and multiple regressions and found that there was an increasing trend in FDI flow, whereas FII flow showed a mixed trend during the study period. They stated that compared to FDI flow, FII flow is less in India. In their study, they found that there was a high degree of positive correlation between FDI and Sensex, and a low degree of correlation between FIIs and Sensex. Hence, they suggested that the government should focus on FDI, which leads to economic progress, and the creation of employment in the country rather than FIIs, and that FDI is most beneficial to the Indian economy compared to other modes of foreign investment. They concluded through multiple regression analysis that FDI and FIIs do not show any impact on BSE Sensex.
- **R. Siva Rama Prasad and Guntupally Lakshmi Vishali** (2017)[9] investigated the role and contribution of FIIs to the Indian capital market. The main objective of this study was to analyze the FII share-holding pattern in various segments of NSE and the relationship between FIIs and the GDP of India. The analysis of data pertaining to sector-wise net FIIs in the NSE, and GDP of India for the past 10 years from 2006–2007 to 2015–2016 found that during 2008–2009, FIIs were unexpectedly pulled out from the Indian capital market due to a change in the investment environment in the United States. They found that Indian promoters hold a major investment share in the NSE, which is followed by FIIs and DIIs. They also stated that most of the investment was flowing into the banking sector as growth compared to the other sectors. The study revealed that there was a continuous growth in the GDP of India, whereas FIIs showed fluctuation during the global recession period (2008–2009).
- **S. Periyasamy and M. Ashok Kumar** (2016)[10] examined the investment behavior of institutional investors (FIIs and DIIs) and its impact on the participation of retail investors in the Indian capital market with the help of the movement of Nifty due to the participation of FIIs and DIIs during the study period from January 2007 to August 2015. Using correlation and regression analysis, they found that there was positive correlation between FIIs and Nifty, a high degree of positive correlation between retail investors and Nifty, a negative correlation between DIIs and retail investors, and a high degree of correlation between

retail investment and Nifty. Furthermore, they found that there was a significant impact of FIIs on retail investors and also that movement of Nifty is positively influenced by FIIs. They concluded that institutional investors are major participants who can move the market in both directions and have an indirect impact on retail investors.

6.2.1 Objectives of the Study

The main objective of the proposed research work was to study the investment pattern of DIIs and its impact on the movement of NSE Nifty returns. The proposed study specifically attempts the following:

1. Evaluate the investment pattern of DIIs in the Indian capital market
2. Examine the impact of DIIs on Nifty returns in the Indian capital market during the pre-recession, recession, and post-recession periods.

6.2.2 Hypotheses of the Study

1. H0: There is no statistically significant change in the investment pattern of DIIs in the Indian capital market during the pre-recession, recession, and post-recession periods.
2. H0: There is no statistically significant influence of the investment pattern of DIIs on the movement of Nifty returns during the pre-recession, recession, and post-recession periods.

6.2.3 Scope of the Study

The proposed study mainly aims at analyzing the investment pattern of FIIs and DIIs along with the investment pattern of retail investors (RIs) in the Indian capital market during the pre-recession, recession, and post-recession periods. In addition, the study also revealed the influence of FIIs, DIIs, and RIs on Nifty returns during the pre-recession, recession, and post-recession periods. Furthermore, the study revealed the relationship among FII, DII, and RI flow of funds in the Indian capital market during the pre-recession, recession, and post-recession periods.

6.3 METHODOLOGY OF THE STUDY

6.3.1 Source of Data

The study is based on the secondary data.[11] The relevant data required for the study is derived from the SEBI annual reports; websites such as moneycontrol, bseindia, nseindia, csdl, and nsdl; and various other published documents on the investment pattern of FIIs, DIIs and retail investors.

6.3.2 Statistical Techniques Used

To evaluate the investment pattern of various investors and their impact on Nifty returns and to understand the relationship among the investors, statistical tools such as mean, standard deviation, coefficient of variation, regression statistics, one-way ANOVA, and coefficient of correlation are used. Linear regression is the most appropriate statistical tool for examining the strength and direction of dependent and independent variables in a statistical relationship.

6.3.3 Period of the Study

The study includes the analysis of secondary data relating to the investment pattern of DIIs and their impact on Nifty returns in the Indian capital market during the financial years 2003–2004 to 2017–2018. The duration of the study is segregated into three phases: the pre-recession period, recession period, and post-recession period. The pre-recession period covers the period of 5 years from 2003–2004 to 2007–2008; the recession period covers the period of 5 years from 2008–2009 to 2012–2013; and the post-recession period covers the period of 5 years from 2013–2014 to 2017–2018.

6.4 RESULTS AND ANALYSIS

6.4.1 Impact of DII Investment on Nifty Returns for the Year 2007–2008

With a view to understand the impact of DII investment pattern on Nifty returns in the Indian capital market for the financial year 2007–2008, the data pertaining to net DII investment and Nifty returns is used. Nifty opening values and Nifty closing values are utilized to compute Nifty returns. The resultant values are presented in Table 6.1.

The data presented in Table 6.1 and Figure 6.1 shows that the month-wise DII investment has an irregular behavior. It shows an increasing trend in some months and a decreasing trend in some other months. The DII investment was Rs. 857.67 crores in April 2007 and increased to Rs. 16,414.7 crores in January 2008, indicating a 19.13 times growth at the end of January 2008. However, during the last 2 months of the 2007–2008 financial year, it showed a decreasing trend compared to January 2008.

The data relating to the Nifty opening index shows an increasing trend in many months of the financial year 2007–2008. The Nifty opening index was 3,820 in April 2007 and increased to 6,136.75 in January 2008, indicating a 1.60 times growth. However, the Nifty opening index during the last 2 months of the 2007–2008 financial year shows a decreasing trend compared to January 2008.

The data relating to the Nifty closing index shows an increasing trend in many months of the financial year 2007–2008. The Nifty closing index was

Table 6.1 Monthly Trends in DII Investment and Nifty Returns for the Year 2007–2008 (Rs. in Crore)

Year	*Month*	*Net DII Investment*	*Nifty Opening*	*Nifty Closing*	*Nifty Returns*
2007–2008	April	857.67	3,820	4,087.9	267.9
	May	3,053.89	4,089.45	4,295.8	206.35
	June	4,560.26	4,296.05	4,318.3	22.25
	July	−587.44	4,318.4	4,528.85	210.45
	August	9,149.24	4,528.85	4,464	−64.85
	September	−4,763.68	4,466.65	5,021.35	554.7
	October	−1,352.18	5,021.5	5,900.65	879.15
	November	8,553.4	5,903.8	5,762.75	−141.05
	December	4,952.15	5,765.45	6,138.6	373.15
	January	16,414.7	6,136.75	5,137.45	−999.3
	February	4,008.31	5,140.6	5,223.5	82.9
	March	2,947.48	5,222.8	4,734.5	−488.3

Source: www.moneycontrol.com, www.nseindia.com, www.sebi.gov.in

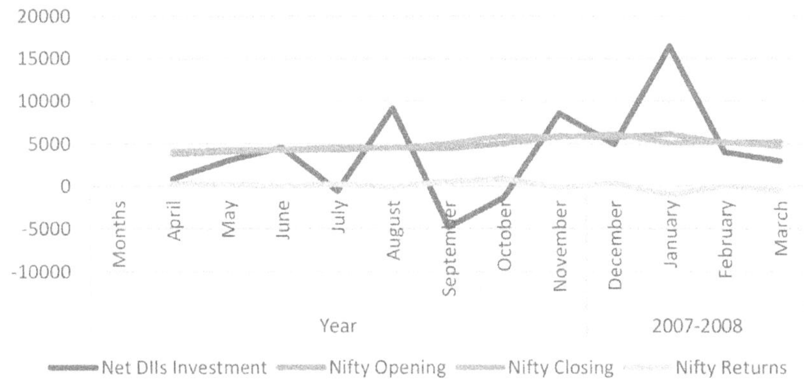

Figure 6.1 Monthly trends in DII investment and Nifty returns for the year 2007–2008 (Rs. in crores).

4,087.9 in April 2007 and increased to 6,138.6 in December 2007, indicating a 1.50 times growth. However, the Nifty closing index during the last 3 months of the 2007–2008 financial year shows a decreasing trend compared to December 2007.

The data relating to Nifty returns shows an irregular behavior of an increasing trend in some months and a decreasing trend in some other months of the financial year 2007–2008. The Nifty returns was 267.9 in April 2007 and increased to 879.15 in October 2007. However, the Nifty returns during the last 5 months of the 2007–2008 financial year shows a decreasing trend compared to October 2007.

6.4.2 Impact of DII Investment on Nifty Returns for the Year 2008–2009

With a view to understanding the impact of the DII investment pattern on Nifty returns in the Indian capital market for the financial year 2008–2009, the data pertaining to net DII investment and Nifty returns is used. Nifty opening values and Nifty closing values are utilized to compute Nifty returns. The resultant values are presented in Table 6.2.

The data presented in Table 6.2 and Figure 6.2 shows that the month-wise DII investment has an irregular behavior. It shows an increasing trend in some months and a decreasing trend in some other months. The

Table 6.2 Monthly Trends in DII Investment and Nifty Returns for the Year 2008–2009 (Rs. in Crores).

Year	*Month*	*Net DII Investment*	*Nifty Opening*	*Nifty Closing*	*Nifty Returns*
2008–2009	April	3,400.97	4,735.65	5,165.9	430.25
	May	8,278.03	5,265.3	4,870.1	−395.2
	June	8,627.96	4,869.25	4,040.55	−828.7
	July	1,527.21	4,039.75	4,332.95	293.2
	August	2,972.36	4,331.6	4,360	28.4
	September	9,212.95	4,356.1	3,921.2	−434.9
	October	10,796.33	3,921.85	2,885.6	−1,036.25
	November	2,874.15	2,885.4	2,755.1	−130.3
	December	1,906.28	2,755.15	2,959.15	204
	January	3,718.14	2,963.3	2,874.8	−88.5
	February	2,774.56	2,872.35	2,763.65	−108.7
	March	3,951.53	2,764.6	3,020.95	256.35

Source: www.moneycontrol.com, www.nseindia.com, www.sebi.gov.in

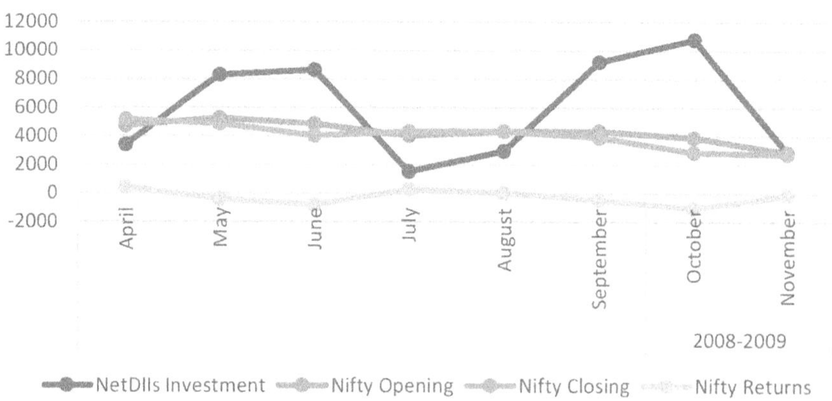

Figure 6.2 Monthly trends in DII investment and Nifty returns for the year 2008–2009 (Rs. in crores).

DII investment was Rs. 3,400.97 crores in April 2008 and increased to Rs. 10,796.33 crores in October 2008, indicating a 3.17 times growth at the end of October 2008. However, during the last 5 months of the 2008–2009 financial year, it showed a decreasing trend compared to October 2008.

The data relating to the Nifty opening index shows a decreasing trend in many months of the financial year 2008–2009. The Nifty opening index was 4,735.65 in April 2008 and increased to 5,265.3 in May 2008, indicating a 1.11 times growth. However, the Nifty opening index during the last 10 months of the 2008–2009 financial year shows a decreasing trend compared to May 2008.

The data relating to the Nifty closing index shows a decreasing trend in many months of the financial year 2008–2009. The Nifty closing index was 5,165.9 in April 2008 and decreased to 2,755.1 in November 2008, indicating a 1.87 times downfall in the Nifty closing index. However, the Nifty closing index during the last 11 months of the 2008–2009 financial year shows a decreasing trend compared to April 2008.

The data relating to Nifty returns shows an irregular increasing trend in some months and a decreasing trend in some other months of the financial year 2008–2009. The Nifty returns was 430.25 in April 2008 and decreased to -1,036.25 in October 2008. However, the Nifty returns during the last 11 months of the 2008–2009 financial year shows a decreasing trend compared to April 2008.

6.4.3 Impact of DII Investment on Nifty Returns for the Year 2009–2010

With a view to understanding the impact of the DII investment pattern on Nifty returns in the Indian capital market for the financial year 2009–2010, the data pertaining to net DII investment Nifty returns is used. Nifty opening values and Nifty closing values are utilized to compute the Nifty returns. The resultant values are presented in Table 6.3.

The data presented in Table 6.3 and Figure 6.3 shows that the month-wise DII investment has an irregular behavior. It shows an increasing trend in some months and a decreasing trend in some other months. The DII investment was Rs.-782.33 crores in April 2009 and increased to Rs. 5,818.98 crores in July 2009, indicating an 8.43 times growth at the end of July 2009. However, during the last 8 months of the 2009–2010 financial year, it showed a decreasing trend compared to July 2009.

The data relating to the Nifty opening index shows an increasing trend in many months of the financial year 2009–2010. The Nifty opening index was 3,023.85 in April 2009 and increased to 5,200.9 in January 2010, indicating a 1.71 times growth. However, the Nifty opening index during the last 2 months of the 2009–2010 financial year shows a decreasing trend compared to January 2010.

The data relating to the Nifty closing index shows an increasing trend in many months of the financial year 2009–2010. The Nifty closing index was

Table 6.3 Monthly Trends in DII Investment and Nifty Returns for the Year 2009–2010 (Rs. in Crores).

Year	*Month*	*Net DII Investment*	*Nifty Opening*	*Nifty Closing*	*Nifty Returns*
2009–2010	April	−782.33	3,023.85	3,473.95	450.1
	May	−82.45	3,478.7	4,448.95	970.25
	June	2,633.78	4,450.4	4,291.1	−159.3
	July	5,818.98	4,292.3	4,636.45	344.15
	August	4,984.71	4,633.8	4,662.1	28.3
	September	696.04	4,662.2	5,083.95	421.75
	October	−47.51	5,087.2	4,711.7	−375.5
	November	2,391.04	4,712.25	5,032.7	320.45
	December	49.67	5,039.7	5,201.05	161.35
	January	1,222.03	5,200.9	4,882.05	−318.85
	February	1,311.95	4,882.05	4,922.3	40.25
	March	−5,135.85	4,935.6	5,249.1	313.5

Source: www.moneycontrol.com, www.nseindia.com, www.sebi.gov.in

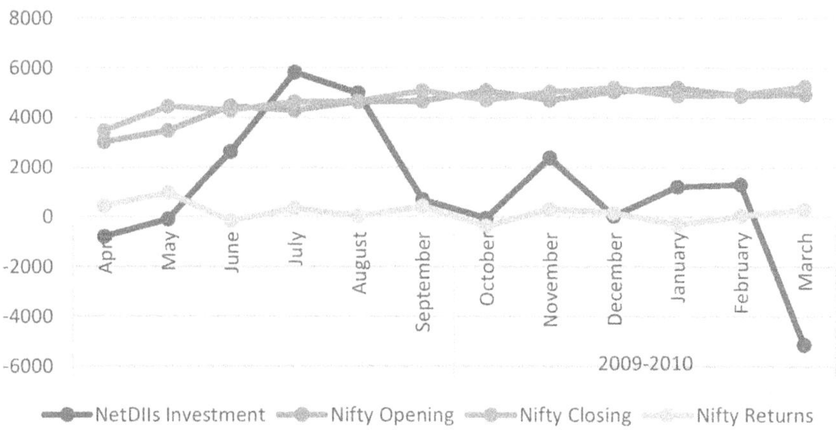

Figure 6.3 Monthly trends in DII investment and Nifty returns for the year 2009–2010 (Rs. in crores).

3,473.95 in April 2009 and increased to 5,249.1 in March 2009, indicating a 1.51 times growth. However, the Nifty closing index during the previous 11 months of the 2009–2010 financial year shows a decreasing trend compared to March 2010.

The data relating to Nifty returns shows an irregular behavior of an increasing trend in some months and a decreasing trend in some other months of the financial year 2009–2010. The Nifty returns was 450.1 in April 2009 and increased to 970 in May 2009. However, the Nifty returns during the last 10 months of the 2009–2010 financial year shows a decreasing trend compared to May 2009.

6.4.4 Impact of DII Investment on Nifty Returns for the Year 2010–2011

With a view to understanding the impact of the DII investment pattern on Nifty returns in the Indian capital market for the financial year 2010–2011, the data pertaining to the net DII investment and Nifty returns is used. Nifty returns opening values and Nifty returns closing values are utilized to compute Nifty returns. The resultant values are presented in Table 6.4.

The data presented in Table 6.4 and Figure 6.4 shows that the month-wise DII investment has an irregular behavior. It shows an increasing trend in

Table 6.4 Monthly Trends in DII Investment and Nifty Returns for the Year 2010–2011 (Rs. in Crores).

Year	*Month*	*Net DII Investment*	*Nifty Opening*	*Nifty Closing*	*Nifty Returns*
2010–2011	April	2,190.66	5,249.2	5,278	28.8
	May	6,361.19	5,278.4	5,086.3	-192.1
	June	−4,777.05	5,086.25	5,312.5	226.25
	July	−6,271.92	5,312.05	5,367.6	55.55
	August	−4,552.68	5,369.55	5,402.4	32.85
	September	−11,866.63	5,403.05	6,029.95	626.9
	October	−11,812.9	6,030.3	6,017.7	−12.6
	November	2,468.82	6,092.3	5,862.7	−229.6
	December	634.99	5,871	6,134.5	263.5
	January	5,237.14	6,177.45	5,505.9	−671.55
	February	6,030.9	5,537.3	5,333.25	−204.05
	March	−37.93	5,382	5,833.75	451.75

Source: www.moneycontrol.com, www.nseindia.com, www.sebi.gov.in

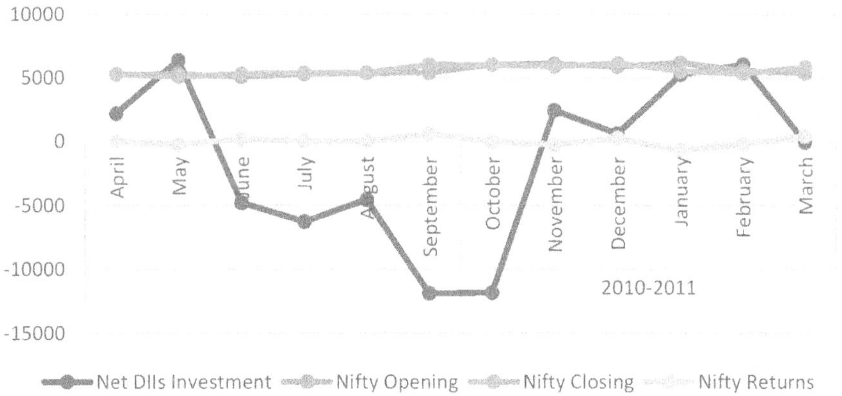

Figure 6.4 Monthly trends in DII investment and Nifty returns for the year 2010–2011 (Rs. in crores).

some months and a decreasing trend in some other months. The DII investment was Rs. 2,190.66 crores in April 2010 and increased to Rs. 6,361.19 crores in May 2010, indicating a 2.90 times growth at the end of May 2010. However, during the last 10 months of the 2010–2011 financial year, it showed a decreasing trend compared to May 2010.

The data relating to the Nifty returns opening index shows an increasing trend in many months of the financial year 2010–2011. The Nifty returns opening index was 5,249.2 in April 2010 and increased to 6,177.45 in January 2011, indicating a 1.17 times growth. However, the Nifty returns opening index during the last 2 months of the 2010–2011 financial year shows a decreasing trend compared to January 2011.

The data relating to the Nifty returns closing index shows an increasing trend in many months of the financial year 2010–2011. The Nifty returns closing index was 5,278 in April 2010 and increased to 6,134.5 in December 2010, indicating a1.16 times growth. However, the Nifty returns closing index during the 3 months of the 2010–2011 financial year shows a decreasing trend compared to December 2011.

The data relating to Nifty returns shows an irregular behavior of an increasing trend in some months and a decreasing trend in some other months of the financial year 2010–2011. The Nifty returns was Rs. 28.8 crores in April 2010 and increased to Rs. 626.9 crores in September 2010. However, the Nifty returns during the last 6 months of the 2010–2011 financial year shows a decreasing trend compared to September 2010.

6.4.5 Impact of DII Investment on Nifty Returns for the Year 2011–2012

With a view to understanding the impact of DII investment patterns on Nifty returns in the Indian capital market for the financial year 2011–2012, data pertaining to the net DII investment and Nifty returns is used. Nifty opening values and Nifty closing values are utilized to compute the Nifty returns. The resultant values are presented in Table 6.5.

The data presented in Table 6.5 and Figure 6.5 shows that the month-wise DII investment exhibited an irregular behavior. It showed an increasing trend in some months and a decreasing trend in others. The DII investment was Rs. –558.27 crores in April 2011 and increased to Rs. 7,991.11 crores in August 2011, indicating a 15.31 times growth by the end of August 2011. However, during the last 7 months of the 2011–2012 financial year, it showed a decreasing trend compared to August 2011.

The data relating to the Nifty opening index shows a decreasing trend in many months of the financial year 2011–2012. The Nifty opening index was 5,835 in April 2011 and decreased to 4,640.20 in January 2012, indicating a 1.25 times fall in the Nifty opening index. However, during the last 2 months of the 2011–2012 financial year, the Nifty opening index showed an increasing trend compared to January 2012.

Table 6.5 Monthly Trends in DII Investment and Nifty Returns for the Year 2011–2012 (Rs. in Crores).

Year	*Month*	*Net DII Investment*	*Nifty Opening*	*Nifty Closing*	*Nifty Returns*
2011–2012	April	−558.27	5,835	5,749.5	−85.5
	May	4,093.21	5,766.9	5,560.15	−206.75
	June	−99.66	5,561.05	5,647.4	86.35
	July	30.07	5,705.75	5,482	−223.75
	August	7,991.11	5,527.5	5,001	−526.5
	September	1,747.09	5,109.8	4,943.25	−166.55
	October	−2,116.64	4,874.4	5,362.6	488.2
	November	5,690.74	5,278.6	4,832.05	−446.55
	December	928.01	4,970.85	4,624.3	−346.55
	January	−6,729.82	4,640.2	5,199.25	559.05
	February	−11,541.38	5,198.15	5,385.2	187.05
	March	−3,490.73	5,366	5,295.55	−70.45

Source: www.moneycontrol.com, www.nseindia.com, www.sebi.gov.in

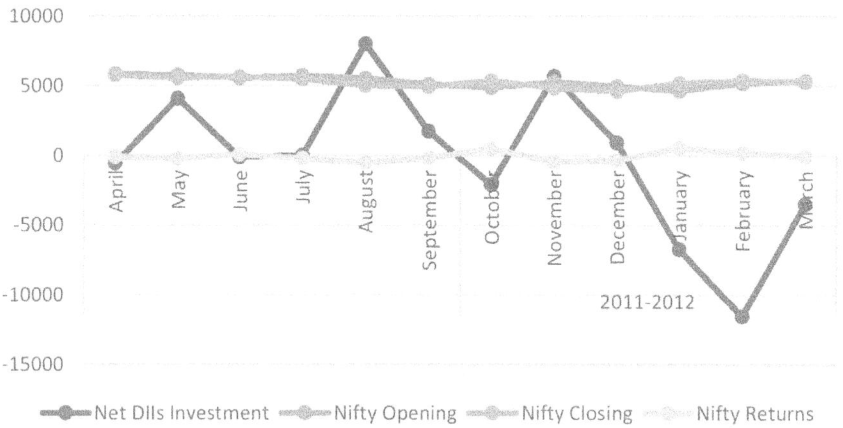

Figure 6.5 Monthly trends in DII investment and Nifty returns for the year 2011–2012 (Rs. in crores).

Similarly, the data relating to the Nifty closing index shows a decreasing trend in many months of the financial year 2011–2012. The Nifty closing index was 5,749.5 in April 2011 and decreased to 4,624.3 in December 2011, indicating a 1.24 times fall in the Nifty closing index. However, during the last 3 months of the 2011–2012 financial year, the Nifty closing index showed an increasing trend compared to December 2011.

The data relating to Nifty returns also displayed an irregular behavior, with an increasing trend in some months and a decreasing trend in others during the financial year 2011–2012. The Nifty returns was −85.5 in April 2011 and increased to 559.05 in January 2012. However, during the last 2 months of the 2011–2012 financial year, the Nifty returns showed a decreasing trend compared to January 2012.

6.4.6 Impact of DII Investment on Nifty Returns for the Year 2012–2013

With a view to understanding the impact of DII investment patterns on Nifty returns in the Indian capital market for the financial year 2012–2013, data pertaining to the net DII investment and Nifty returns are used. Nifty opening values and Nifty closing values are utilized to compute the Nifty returns. The resultant values are presented in Table 6.6.

The data presented in Table 6.6 and Figure 6.6 shows that the month-wise DII investment exhibited an irregular behavior. It showed an increasing trend in some months and a decreasing trend in others. The DII investment was Rs. 783.01 crores in April 2012 and increased to Rs. 1,102.75 crores in June 2012, indicating a 1.40 times growth by the end of June 2012. However, during the last 9 months of the 2012–2013 financial year, it showed a decreasing trend compared to June 2012.

The data relating to the Nifty opening index shows an increasing trend in many months of the financial year 2012–2013. The Nifty opening index was 5,296.35 in April 2012 and increased to 6,040.95 in February 2013, indicating a 1.14 times growth. However, during the last month of the 2012–2013 financial year, the Nifty opening index showed a decreasing trend compared to February 2013.

The data relating to the Nifty closing index also shows an increasing trend in many months of the financial year 2012–2013. The Nifty closing index was 5,248.15 in April 2012 and increased to 6,034.75 in January

Table 6.6 Monthly Trends in DII Investment and Nifty Returns for the Year 2012–2013 (Rs. in Crores).

Year	*Month*	*Net DII Investment*	*Nifty Opening*	*Nifty Closing*	*Nifty Returns*
2012–2013	April	783.01	5,296.35	5,248.15	−48.2
	May	707.22	5,254.3	4,924.25	−330.05
	June	1,102.75	4,910.85	5,278.9	368.05
	July	−5,270.67	5,283.85	5,229	−54.85
	August	−4,408.63	5,220.7	5,258.5	37.8
	September	−9,152.2	5,276.5	5,703.3	426.8
	October	−4,695.11	5,704.75	5,619.7	−85.05
	November	−4,201.42	5,609.85	5,879.85	270
	December	−8,903.11	5,878.25	5,905.1	26.85
	January	−16,207.32	5,937.65	6,034.75	97.1
	February	−8,818.5	6,040.95	5,693.05	−347.9
	March	−7,872.45	5,702.45	5,682.55	−19.9

Source: www.moneycontrol.com, www.nseindia.com, www.sebi.gov.in

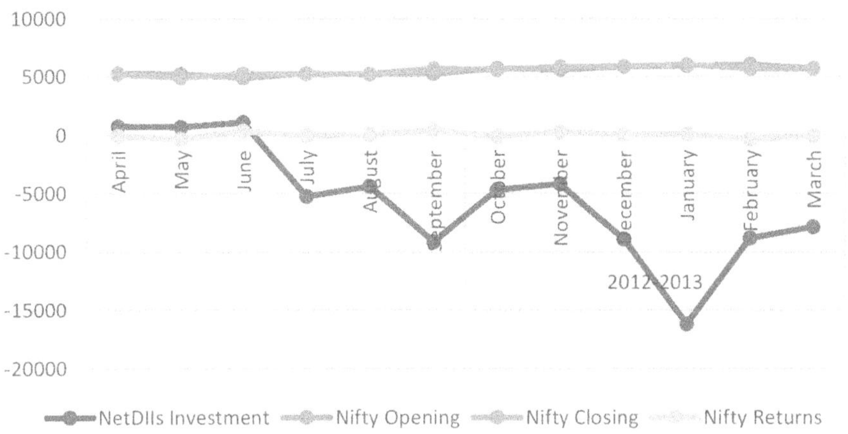

Figure 6.6 Monthly trends in DII investment and Nifty returns for the year 2012–2013 (Rs. in crores).

2013, indicating a 1.14 times growth. However, during the last 2 months of the 2012–2013 financial year, the Nifty closing index showed a decreasing trend compared to January 2013.

The data relating to Nifty returns displays an irregular behavior, with an increasing trend in some months and a decreasing trend in others during the financial year 2012–2013. The Nifty returns was –48.2 in April 2012 and increased to 426.80 in September 2012. However, during the last 6 months of the 2012–2013 financial year, the Nifty returns showed a decreasing trend compared to September 2012.

6.4.7 Impact of DII Investment on Nifty Returns for the Year 2013–2014

With a view to understanding the impact of the DII investment pattern on Nifty returns in the Indian capital market for the financial year 2013–2014, the data pertaining to net DII investment and Nifty returns is used. Nifty opening values and Nifty closing values are utilized to compute the Nifty returns. The resultant values are presented in Table 6.7.

The data presented in Table 6.7 and Figure 6.7 shows that the month-wise DII investment has an irregular behavior. It shows an increasing trend in some months and a decreasing trend in some other months.

The DII investment was Rs. –2,998.27 crores in April 2013 and increased to Rs. 8,427.42 crores in June 2013, indicating a 3.81 times growth at the end of June 2013. However, during the previous months of the 2013–2014 financial year, it showed a decreasing trend compared to June 2013.

The data relating to the Nifty opening index shows an increasing trend in many months of the financial year 2013–2014. The Nifty opening index was

Table 6.7 Monthly Trends in DII Investment and Nifty Returns for the Year 2013–2014 (Rs. in Crores).

Year	*Month*	*Net DII Investment*	*Nifty Opening*	*Nifty Closing*	*Nifty Returns*
2013–2014	April	−2,998.27	5,697.35	5,930.2	232.85
	May	−12,052.43	5,911.4	5,985.95	74.55
	June	8,427.42	5,997.35	5,842.2	−155.15
	July	−1,540.85	5,834.1	5,742	−92.1
	August	6,284.72	5,776.9	5,471.8	−305.1
	September	−9,130.07	5,480.25	5,735.3	255.05
	October	−12,410.71	5,756.1	6,299.15	543.05
	November	−9,147.17	6,289.75	6,176.1	−113.65
	December	−7,586.06	6,171.15	6,304	132.85
	January	−1,455.47	6,323.8	6,089.5	−234.3
	February	667.69	6,058.8	6,276.95	218.15
	March	−13,130.77	6,264.35	6,704.2	439.85

Source: www.moneycontrol.com, www.nseindia.com, www.sebi.gov.in

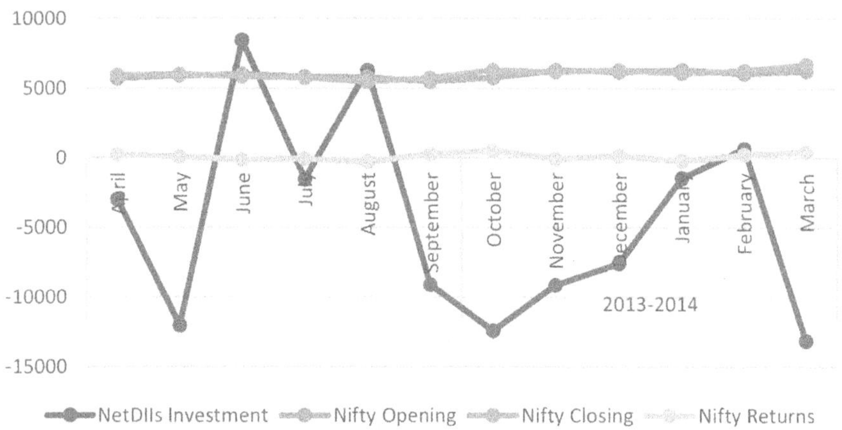

Figure 6.7 Monthly trends in DII investment and Nifty returns for the Year 2013–2014 (Rs. in crores).

5,697.35 in April 2013 and increased to 6,323.8 in January 2014, indicating a 1.10 times growth. However, the Nifty opening index during the last 2 months of the 2013–2014 financial year shows a decreasing trend compared to January 2014.

The data relating to the Nifty closing index shows an increasing trend in many months of the financial year 2013–2014. The Nifty closing index was 5,930.2 in April 2013 and increased to 6,704 in March 2014, indicating a 1.13 times growth. However, the Nifty closing index during the previous 11 months of the 2013–2014 financial year shows a decreasing trend compared to March 2014.

The data relating to Nifty returns shows an irregular behavior, with an increasing trend in some months and a decreasing trend in some other months of the financial year 2013–2014. The Nifty returns was 232.85 in April 2013 and increased to 543.05 in October 2013. However, the Nifty returns during the last 5 months of the 2013–2014 financial year shows a decreasing trend compared to October 2013.

6.4.8 Impact of DII Investment on Nifty Returns for the Year 2014–2015

With a view to understanding the impact of the DII investment pattern on Nifty returns in the Indian capital market for the financial year 2014–2015, the data pertaining to net DII investment and Nifty returns is used. Nifty opening values and Nifty closing values are utilized to compute Nifty returns. The resultant values are presented in Table 6.8.

The data presented in Table 6.8 and Figure 6.8 shows that the month-wise DII investment exhibited an irregular behavior. It displayed an increasing trend in some months and a decreasing trend in others. The DII investment was Rs. -6,859.04 crores in April 2014 and increased to Rs. 5,955.79 crores in December 2014, indicating a 1.86 times growth by the end of December 2014. However, during the last 3 months of the 2014–2015 financial year, it showed a decreasing trend compared to December 2014.

The data relating to the Nifty opening index indicates an increasing trend in many months of the financial year 2014–2015. The Nifty opening index was 6,729.50 in April 2014 and increased to 8,953.85 in March 2015, indicating a 1.33 times growth. However, the Nifty opening index during the previous 11 months of the 2014–2015 financial year showed a decreasing trend compared to March 2015.

Table 6.8 Monthly Trends in DII Investment and Nifty Returns for the Year 2014–2015 (Rs. in Crores).

Year	*Month*	*Net DII Investment*	*Nifty Opening*	*Nifty Closing*	*Nifty Returns*
2014–2015	April	−6,859.04	6,729.5	6,696.4	−33.1
	May	−4,136.04	6,709.95	7,229.95	520
	June	−4,412.68	7,264.05	7,611.35	347.3
	July	−3,216.53	7,629	7,721.3	92.3
	August	2,335.48	7,662.5	7,954.35	291.85
	September	−1,136.99	7,990.35	7,964.8	−25.55
	October	4,102.87	7,960.5	8,322.2	361.7
	November	−7,271.34	8,348.15	8,588.25	240.1
	December	5,955.79	8,605.1	8,282.7	−322.4
	January	−6,531.75	8,272.8	8,808.9	536.1
	February	1,711.88	8,802.5	8,901.85	99.35
	March	194.76	8,953.85	8,491	−462.85

Source: www.moneycontrol.com, www.nseindia.com, www.sebi.gov.in

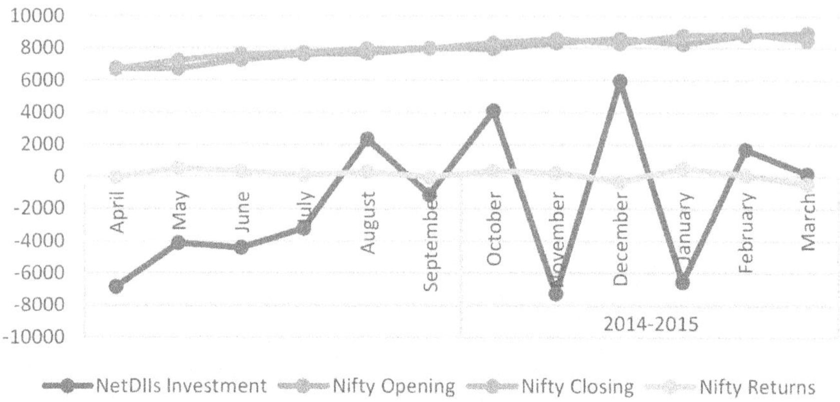

Figure 6.8 Monthly trends in DII investment and Nifty returns for the year 2014–2015 (Rs in crores),

The data relating to the Nifty closing index also shows an increasing trend in many months of the financial year 2014–2015. The Nifty closing index was 6,696.40 in April 2014 and increased to 8,901.85 in February 2015, indicating a 1.32 times growth. However, the Nifty closing index during the last month of the 2014–2015 financial year showed a decreasing trend compared to February 2015.

The data relating to Nifty returns show an irregular pattern with an increasing trend in some months and a decreasing trend in others during the financial year 2014–2015. The Nifty returns was –33.1 in April 2014 and increased to 536.1 in January 2015. However, the Nifty returns during the last 2 months of the 2014–2015 financial year showed a decreasing trend compared to January 2015.

6.4.9 Impact of DII Investment on Nifty Returns for the Year 2015–2016

With a view to understanding the impact of DII investment on Nifty returns in the Indian capital market for the financial year 2015–2016, the data pertaining to net DII investment and Nifty returns is used. Nifty opening values and Nifty closing values are utilized to compute the Nifty returns. The resultant values are presented in Table 6.9.

The data presented in Table 6.9 and Figure 6.9 shows that the month-wise DII investment exhibited an irregular behavior. It showed an increasing trend in some months and a decreasing trend in others. The DII investment was Rs. 11,511.46 crores in April 2015 and increased to Rs. 16,437.53 crores in August 2015, indicating a 1.42 times growth by the end of August 2015. However, during the last 7 months of the financial year 2015–2016, it showed a decreasing trend compared to August 2015.

Table 6.9 Monthly Trends in DII Investment and Nifty Returns for the Year 2015–2016 (Rs in Crores)

Year	*Month*	*Net DII Investment*	*Nifty Opening*	*Nifty Closing*	*Nifty Returns*
2015–2016	April	11,511.46	8,483.7	8,181.5	−302.2
	May	8,582.26	8,230.05	8,433.65	203.6
	June	12,026.55	8,417.25	8,368.5	−48.75
	July	72.31	8,376.25	8,532.85	156.6
	August	16,437.53	8,510.65	7,971.3	−539.35
	September	10,273.06	7,907.95	7,948.9	40.95
	October	−1,518.86	7,992.05	8,065.8	73.75
	November	8,499.98	8,054.55	7,935.25	−119.3
	December	6,327.64	7,958.15	7,946.35	−11.8
	January	12,874.9	7,938.45	7,563.55	−374.9
	February	10,491.61	7,589.5	6,987.05	−602.45
	March	−16,891.9	7,038.25	7,738.4	700.15

Source: www.moneycontrol.com, www.nseindia.com, www.sebi.gov.in

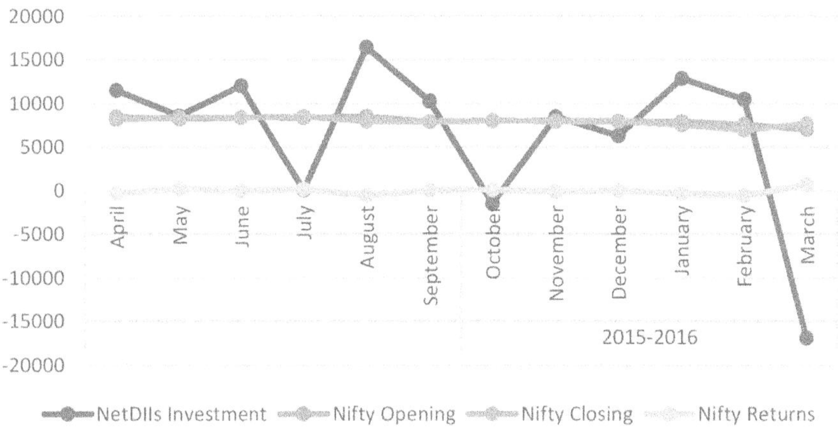

Figure 6.9 Monthly trends in DII investment and Nifty returns for the year 2015–2016 (Rs. in crores).

The data relating to the Nifty opening index indicates a decreasing trend in many months of the financial year 2015–2016. The Nifty opening index was 8,483.7 in April 2015 and slightly increased to 8,510.65 in August 2015, reflecting only a 1.00 times growth. However, during the last 7 months of the year, the Nifty opening index showed a decreasing trend compared to August 2015.

The data on the Nifty closing index reveals an increasing trend in several months of the financial year 2015–2016. The Nifty opening index was 8,181.5 in April 2015 and increased to 8,532.85 in July 2015, indicating a

1.04 times growth. However, during the last 8 months of the financial year, the closing index showed a declining trend compared to July 2015.

The Nifty returns also demonstrated an irregular behavior, with increases in some months and decreases in others during the financial year 2015–2016. The Nifty returns was -302.2 in April 2015 and increased significantly to 700.15 in December 2015. However, during the remaining months leading up to March 2016, the Nifty returns showed a decreasing trend compared to December 2015.

6.4.10 Impact of DII Investment on Nifty Returns for the Year 2016–2017

With a view to understanding the impact of the DII investment pattern on Nifty returns in the Indian capital market for the financial year 2016–2017, the data pertaining to net DII investment and Nifty returns is used. Nifty opening values and Nifty closing values are utilized to compute the Nifty returns. The resultant values are presented in Table 6.10.

The data presented in Table 6.10 and Figure 6.10 shows that the month-wise DII investment had an irregular behavior. It showed an increasing trend in some months and a decreasing trend in others. The DII investment was Rs. -2554.36 crores in April 2016 and increased to Rs. 18,277.03 crores in November 2016, indicating an 8.15 times growth by the end of November 2016. However, during the last 4 months of the 2016–2017 financial year, it showed a decreasing trend compared to November 2016.

The data relating to the Nifty opening index shows an increasing trend in many months of the financial year 2016–2017. The Nifty opening index was 7,718.05 in April 2016 and increased to 8,904.4 in March 2017, indicating

Table 6.10 Monthly Trends in DII Investment and Nifty Returns for the Year 2016–2017 (Rs. in Crores).

Year	*Month*	*Net DII Investment*	*Nifty Opening*	*Nifty Closing*	*Nifty Returns*
2016–2017	April	−2,554.36	7,718.05	7,849.8	131.75
	May	6,758.08	7,822.7	8,160.1	337.4
	June	−2,173.55	8,179.2	8,287.75	108.55
	July	−6,054.6	8,313.05	8,638.5	325.45
	August	−4,406.31	8,654.3	8,786.2	131.9
	September	1,999.22	8,793.6	8,611.15	-182.45
	October	7,906.35	8,666.15	8,625.7	−40.45
	November	18,277.03	8,653.15	8,224.5	−428.65
	December	9,136.09	8,244	8,185.8	−58.2
	January	4,503.94	8,210.1	8,561.3	351.2
	February	935.26	8,570.35	8,879.6	309.25
	March	−4,395.61	8,904.4	9,173.75	269.35

Source: www.moneycontrol.com, www.nseindia.com, www.sebi.gov.in

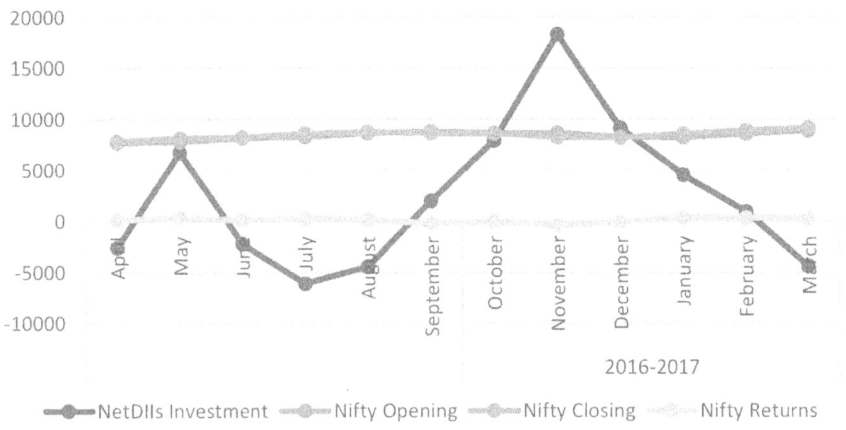

Figure 6.10 Monthly trends in DII investment and Nifty returns for the year 2016–2017 (Rs. in crores).

a 1.15 times growth. However, during the previous 11 months of the 2016–2017 financial year, the Nifty opening index showed a decreasing trend compared to March 2017.

The data relating to the Nifty closing index also shows an increasing trend in many months of the financial year 2016–2017. The Nifty closing index was 7,849.8 in April 2016 and increased to 9,173.75 in March 2017, indicating a 1.16 times growth. However, during the previous 11 months of the financial year, the Nifty closing index showed a decreasing trend compared to March 2017.

The data relating to Nifty returns shows an irregular behavior, with an increasing trend in some months and a decreasing trend in others during the financial year 2016–2017. The Nifty returns was 131.75 in April 2016 and increased to 351.2 in January 2017. However, in the last 2 months of the financial year, the Nifty returns showed a decreasing trend compared to January 2017.

6.4.11 Impact of DII Investment on Nifty Returns for the Year 2017–2018

With a view to understanding the impact of DII investment on Nifty returns in the Indian capital market for the financial year 2017–2018, the data pertaining to net DII investment and Nifty returns is used. Nifty opening values and Nifty closing values are utilized to compute the Nifty returns. The resultant values are presented in Table 6.11.

The data presented in Table 6.11 and Figure 6.11 shows that the month-wise DII investment has an irregular behavior. It showed an increasing trend in some months and a decreasing trend in some other months. The

Table 6.11 Monthly Trends in DII Investment and Nifty Returns for the Year 2017–2018 (Rs. in Crores).

Year	*Month*	*Net DII Investment*	*Nifty Opening*	*Nifty Closing*	*Nifty Returns*
2017–2018	April	9,247.43	9,220.6	9,304.05	83.45
	May	4,277.06	9,339.85	9,621.25	281.4
	June	6,676.11	9,603.55	9,520.9	−82.65
	July	4,786.37	9,587.95	10,077.1	489.15
	August	16,205.22	10,101.05	9,917.9	−183.15
	September	21,025.53	9,937.65	9,788.6	−149.05
	October	10,090.91	9,893.3	10,335.3	442
	November	9,243.21	10,390.35	10,226.55	−163.8
	December	8,142.88	10,263.7	10,530.7	267
	January	398.73	10,531.7	11,027.7	496
	February	17,813.01	11,044.55	10,492.85	−551.7
	March	6,693.91	10,479.95	10,113.7	−366.25

Source: www.moneycontrol.com, www.nseindia.com, www.sebi.gov.in

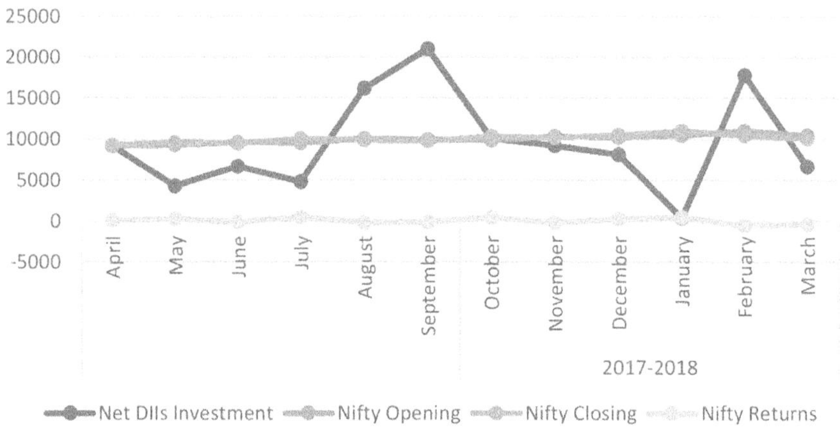

Figure 6.11 Monthly trends in DII investment and Nifty returns.

DII investment was Rs. 9,247.43 crores in April 2017 and increased to Rs. 21,025.53 crores in September 2017, indicating a 2.27 times growth at the end of September 2017. However, during the last 6 months of the 2017–2018 financial year, it showed a decreasing trend compared to September 2017.

The data relating to the Nifty opening index shows an increasing trend in many months of the financial year 2017–2018. The Nifty opening index was 9,220.6 in April 2017 and increased to 11,044.55 in February 2018, indicating a 1.19 times growth. However, the Nifty opening index during the last 1 month of the 2017–2018 financial year shows a decreasing trend compared to February 2018.

The data relating to the Nifty closing index shows an increasing trend in many months of the financial year 2017–2018. The Nifty closing index was 9,304.05 in April 2017 and increased to 11,027.7 in January 2018, indicating a 1.18 times growth. However, the Nifty closing index during the last months of the 2017–2018 financial year shows a decreasing trend compared to January 2018.

The data relating to Nifty returns show an irregular behavior of an increasing trend in some months and a decreasing trend in some other months of the financial year 2017–2018. The Nifty returns was 83.45 in April 2017 and increased to 496 in January 2018. However, the Nifty returns during the last 2 months of the 2017–2018 financial year shows a decreasing trend compared to January 2018.

Table 6.12 and Figure 6.12 reveal that the total net DII investment for the pre-recession period was Rs. 47,793.8 crores. There was a growth of 5.11% in net DII investment in 2008–2009 compared to the previous year.

During the recession period, there was more outflow of DII investment than inflow of capital. The net DII investment in 2008–2009 was Rs. 50,235.23 crores, which declined to negative Rs. 66,936.40 crores, reflecting a significant withdrawal of domestic investment from the Indian capital market. However, the highest growth and growth percentage during the recession period were witnessed in the year 2011–2012.

In the post-recession period, some fresh domestic capital inflow was observed due to favorable economic conditions. Until 2014–2015, there was more withdrawal than investment, as disclosed in the table. High growth and growth percentage were recorded in the year 2015–2016 during the post-recession period. By the end of 2017–2018, the total DII investment

Table 6.12 Trends in Flow of DII Investments during 2007–2008 to 2017–2018 (Rs. in Crores).

Period	*Year*	*Total Net DII Investment (Rs.)*	*Growth (Rs.)*	*Growth in Percentage*
Pre-Recession	2007–2008	47,793.8	-	-
Recession	2008–2009	50,235.23	2,441.43	5.11
	2009–2010	13,060.06	−37,175.17	−74.00
	2010–2011	−16,395.40	−29,455.46	−225.54
	2011–2012	−4,056.27	12,339.13	75.26
	2012–2013	−66,936.40	−62,880.13	−1550.20
Post-Recession	2013–2014	−54,072.00	12,864.40	19.22
	2014–2015	−19,263.59	34,808.41	64.37
	2015–2016	78,686.54	97,950.13	508.47
	2016–2017	29,931.54	−48,755.00	−61.96
	2017–2018	114,600.37	84,668.83	282.87

Source: www.moneycontrol.com, www.nseindia.com

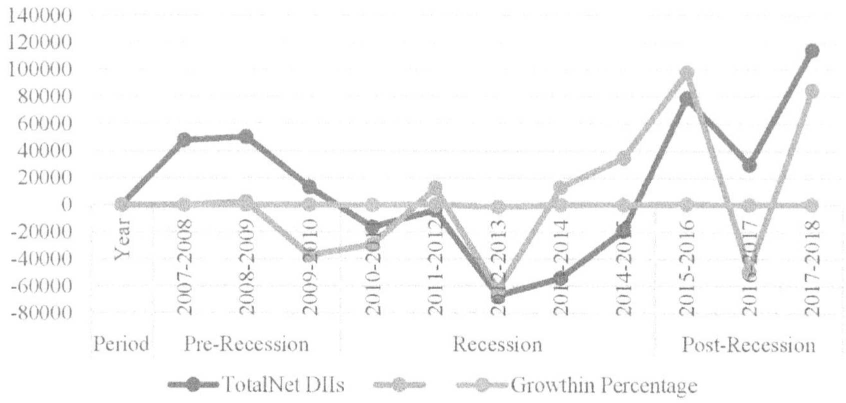

Figure 6.12 Trends in flow of DII investments during 2007–2008 to 2017–2018 (Rs. in crores).

reached Rs. 1,14,600.37 crores, representing a 282.87% growth from the DII investment balance of Rs. 29,931.54 crores in 2016–2017.

6.4.12 Overall Impact of DII Investment on Nifty Returns

With a view to examine the overall impact of DII investment on Nifty returns, the regression results during the pre-recession, recession, and post-recession periods are presented in Table 6.13.

An examination of data presented in Table 6.13 shows that the R value is 0.811 in the year 2007–2008 during the pre-recession period, indicating a stronger relationship between DII investment and Nifty returns.

The R-square value during the pre-recession period 2007–2008 is 0.658, which indicates that the independent variable DII investment explains around 65.8% variation in the dependent variable Nifty returns. Examination of the p value reveals that there was a significant influence of DII investment on Nifty returns during 2007–2008.

The β value during the pre-recession period is −0.071. It implies that the DII investment increases by one unit and Nifty returns decreases by a respective value (β values in the year).

An examination of the data presented in Table 6.13 shows that the R value in the year 2008–2009 varied between 0.08 and 0.878 during the pre-recession period, indicating a stronger relationship between DII investment and Nifty returns, whereas a weak relationship can be seen in the year 2012–2013 as the R value was 0.08.

The R-squared values during the recession period varied between 0.006 during 2012–2013 and 0.771 during the period 2008–2009, indicating that

Table 6.13 Summary of the Relationship between DII Investment and Nifty Returns during 2007–2008 to 2017–2018.

Period	*Year*	R *Value*	R-*Square*	p *Value at 5% Level of Sig.*	β *Value*
Pre-recession	2007–2008	0.811	0.658	0.001	−0.071
Recession	2008–2009	0.878	0.771	0.000	−0.122
	2009–2010	0.169	0.028	0.60	−0.022
	2010–2011	0.604	0.365	0.037	−0.032
	2011–2012	0.76	0.578	0.004	−0.049
	2012–2013	0.08	0.006	0.804	−0.004
Post-recession	2013–2014	0.68	0.463	0.015	−0.025
	2014–2015	0.392	0.154	0.208	−0.027
	2015–2016	0.835	0.698	0.001	−0.033
	2016–2017	0.646	0.417	0.023	−0.022
	2017–2018	0.636	0.404	0.026	−0.037

the independent variable DII investment explains around 6.00–77.1% variation in the dependent variable Nifty returns.

An examination of the *p* value during the recession period reveals that there was a significant influence of DII investment on Nifty returns only during the periods 2008–2009, 2010–2011, and 2011–2012. There was a consistent increase in the *β* values during the recession period from –0.022 in 2009–2010 to –0.049 in 2011–2012, except during the period 2008–2009, where a slight increase was witnessed (i.e., –0.122) compared to the following 3 years. A substantial decrease was also observed (i.e., –0.004) compared to the previous 4 years. This shows that during the recession period, the influence of DII investment on Nifty returns followed an increasing trend.

An examination of the data presented in Table 6.13 shows that the *R* value varied between 0.392 and 0.835 during the post-recession period. The *R* value of 0.835 in the year 2015–2016 indicates a stronger relationship between DII investment and Nifty returns, whereas a weak relationship can be seen in the year 2014–2015 as the *R* value was 0.392.

The *R*-square values during the post-recession period varied between 0.154 in 2014–2015 and 0.698 in 2015–2016, indicating that the independent variable, DII investment, explains around 15.4–69.8% of the variation in the dependent variable, Nifty returns.

An examination of the *p* value during the post-recession period revealed that there was a significant influence of DII investment on Nifty returns, except during the period 2014–2015.

There was a consistent increase in the *β* values during the post-recession period, from -0.025 in 2013–2014 to 0.037 in 2017–2018, except during the period 2016–2017, which showed a decreasing trend from -0.033 to -0.022. This shows that during the post-recession period, DII investment influence on Nifty returns followed an increasing trend overall. The value of

β implies that if DII investment increases by one unit, Nifty returns increases by the respective β value in that year. The β value was highest in 2017–2018 and lowest in 2016–2017, explaining the variability in the model's relationship.

NOTES

1. Debasis Mohanty Vikram and Archa Agrawal (2025). An Empirical Study on the Impact of FII and DII on Volatility, Leverage and Long-Term Returns of the Indian Stock Index. *Asian Journal of Economics, Business and Accounting*, 25(4), 131–137.
2. S. Sunkara, P. K. Sinha, N. Harish, V. L. Suneetha, K. T. R. Reddy, and K. L. Kumar (2025). Testing the Impact of FII Activity and DII Activity on Benchmark Index Nifty 50 Concerning India. *International Journal of Environmental Sciences*, 11(16s), 1842–1844.
3. P. Swapna, John Imran Sandeep and M. Kameswara Rao (2024, Jul–Dec). The Impact of DIIs on Indian Banking Sector: Growth Patterns and Future Outlook. *Library Progress International* 44(3), 20169–20182.
4. M. Rathod, S. Kumari, P. Kujur, R. Bisht, S. Paul, and A. Debnath (2024). Impact of DIIs and FIIs on the Indian Stock Market. *African Journal of Biomedical Research*, 27(4S), 12200–12204.
5. M. Gandhi and Vivek Ayre (2024, Mar.–Apr.). A Study on Impact of FII and DII on Indian Stock Market (BSE Sensex). *International Journal of Computational Engineering Research (IJCER)*, ISSN (e): 2250–3005, 14(2).
6. P. M. Mathew and N. K. John (2019). A Study on the Impact of Institutional Investors and Mutual Funds on Indian Stock Market. *Journal of Emerging Technologies and Innovative Research*, 6(3), 333–341.
7. S. A. Atif Salar and Mohd Shamim (2017). Trading Behaviour of Domestic Institutional Investors and Volatility of Indian Stock Market. *Amity Journal of Finance*, 2(1), 47–55.
8. Tanu Aggarwal and Priya Solomon (2017). Trends and Patterns of FDI and FII in Indian—Implication for the Future. *Global Journal of Management and Business Research (B)*, XVII, Version 1.
9. R. Shiva Rama Prasad and Guntupally Lakshmi Vishali (2017). An Empirical Study on FII Investment Pattern in Indian Capital Market. *International Journal of Applied Research*, 3(12), 362–367.
10. S. Periyasamy and M. Ashok Kumar (2016, February). A Study on the Impact of Institutional Investors Contribution on the Indian Stock Market with Reference to Nifty. *ZENITH International Journal of Multidisciplinary Research*, 6(2), 61–67.
11. All time series are stationary at first difference, i.e., I(1). After testing them with the augmented dickey fuller test unit root test, the results obtained are not reported in the paper. However, the result of the unit root test is provided upon request.

Chapter 7

A New Approach for Finding a Solution to the Fuzzy Transportation Problem under Fuzzy Environments

Ananya Mishra and Siva Prasad Behera

7.1 INTRODUCTION

Transportation problems are a specific type of linear programming problem (LPP) that involves finding the most cost-effective way to distribute a single product from various suppliers to multiple demand points. Effective algorithms are available for solving transportation problems when supply, demand, and transportation costs are known with precision. In real-world scenarios, decision-makers frequently encounter incomplete or imprecise information regarding various parameters. Traditional transportation methods, which rely on precise numbers, may not suffice. To address this, efficient methods like interval or fuzzy numbers can be used to handle uncertainties and optimize solutions.

Bellman and Zadeh [1] laid the groundwork for fuzzy logic. Chanas later extended this work to transportation problems, employing a parametric method to tackle fuzzy transportation issues. According to Hillier and Lieberman [2], transportation problem is an exceptional use of LPP that focuses on determining the ideal amount of produced items supplied from various sources to different destination locations so as to minimize the overall cost of transportation. The fuzzy transportation problem has been the subject of numerous studies. Zadeh [3] was the first to suggest the idea of fuzzy sets. Nagoor Gani and Abdul Rajak [4] developed a fuzzy solution to a fuzzy transportation problem using a parametric technique. The fuzzy linear programming (LP) method was developed by Zimmerman [5]. A straightforward approach is to use the current fuzzy LP techniques [6–9] to solve the fuzzy transportation problem as it is fundamentally a LPP. Bharati [10] explored interval-valued intuitionistic fuzzy sets (IVIFSs). She proposed a new ranking function to tackle uncertainty and hesitation in real-life problems, such as transportation costs in a fuzzy transportation problem. R. Srinivasan [11, 12] proposed a new algorithm to solve the fuzzy transportation problem. Wieczkowski [13] proposed a comprehensive technique for fuzzy multi-criteria decision-making problems. Pooja et al. [14] proposed a new algorithm for solving fuzzy transportation problems using

DOI: 10.1201/9781003642091-7

trapezoidal fuzzy numbers and ranking methodology, validated through a numerical example. The transportation problem is found globally in solving certain real-world problems [15]. The sources and the destinations are represented in terms of fuzzy trapezoidal numbers in two ways. Nagoor Gani and Abdul Rajak [4] developed a fuzzy solution to a fuzzy transportation problem using a parametric technique. Dhouib [16] proposed a column-row heuristic method, Dhouib-Matrix-TP1, enhanced with a robust ranking function, to efficiently solve transportation problems in a trapezoidal fuzzy environment, validated through numerical experiments. Some recent studies [17–20] proposed a novel approach to solve fuzzy transportation issues. In our proposed method, we tackle the fuzzy transportation problem by treating each source, each destination, and each transportation cost as a fuzzy number.

This study addresses a significant gap in existing research by developing a comprehensive framework that combines trapezoidal fuzzy numbers with a pioneering solution method, specifically designed to tackle the complexities of transportation problems amid uncertainty. Our innovative approach harnesses the power of trapezoidal fuzzy numbers to accurately represent ambiguous transportation data, yielding more effective solutions. By doing so, we aim to improve decision-making in transportation logistics, providing more precise and robust outcomes that account for real-world fuzziness. This research bridges the divide between theoretical fuzzy set advancements and practical transportation problem-solving needs, ultimately contributing to more robust and efficient supply chain management systems.

Section 7.2 of this study recalls some of the preliminary concepts and definitions required to comprehend our results. Section 7.3 presents a novel technique for resolving the fuzzy transportation problem. Section 7.4 discusses a numerical example. In Section 7.5, a comparative study is conducted between the different values of the optimal solution with a graph present. In Section 7.6, we conclude our findings.

7.2 PRELIMINARIES

Some basic definitions defined here are used to compute the design of the new algorithm.

7.2.1 Definition

Here, the fuzzy number $\tilde{B}$ in the set S said to represent a fuzzy trapezoidal number with a membership function $\mu_{\tilde{B}} : S \rightarrow [0,1]$ is defined as follows. A trapezoidal fuzzy number defined by four points (m,n,o,p) is presented in Figure 7.1.

The fuzzy trapezoidal number is denoted as $\tilde{B} = (m, n, o, p)$,

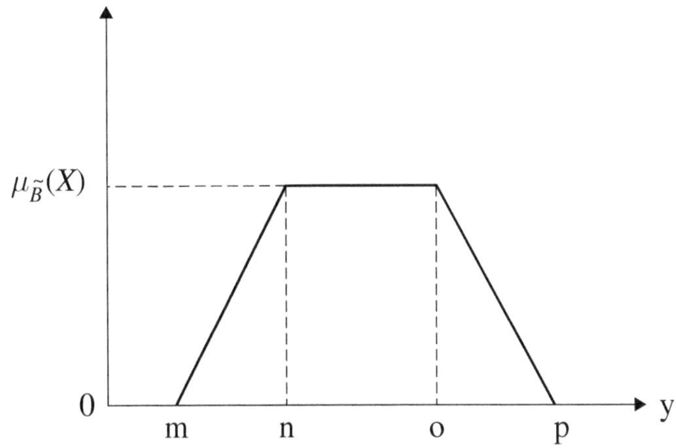

Figure 7.1 Graphical representation of a trapezoidal fuzzy number.

where

m = lower bound, n = lower core value, o = upper core value, p = upper bound.

$$\mu_{\tilde{B}}(X) = \begin{cases} 0, & \text{if } x \le m \\ \dfrac{x-m}{n-m} & \text{if } m < x \le n \\ 1, & \text{if } n \le x \le o \\ \dfrac{o-x}{p-o} & \text{if } o < x \le p \\ 0, & \text{if } x \ge p \end{cases}$$

Let's say a trapezoidal fuzzy number is defined as (3, 5, 7, 9). This means the following:

- The membership value is 0 for any value less than 3 or greater than 9.
- The membership value is 1 for any value between 5 and 7 (inclusive).
- The membership value increases from 0 to 1 between 3 and 5, and decreases from 1 to 0 between 7 and 9.

7.2.2 Definition

Let $\tilde{B} = (m, n, o, p)$ be a trapezoidal fuzzy number. The center of gravity approach is used to evaluate the crisp value corresponding to each fuzzy trapezoidal number. $\tilde{B} = (m, n, o, p)$.

7.2.3 Transportation Problem

Mathematically, a transportation problem is defined as follows.

Let $D_{kl} = (m_{kl}, n_{kl}, o_{kl}, p_{kl})$ be the trapezoidal fuzzy transportation cost from source k to destination l,

and $\tilde{X}_{kl}$ be the fuzzy quantity transported from source k to destination l.

Objective Function:

Minimize

$$Z = \sum_{k=1}^{u} \sum_{l=1}^{v} D_{kl} X_{kl}$$

Subject to the constraints

$$\sum_{l=1}^{v} X_{kl} = p_k \qquad k = 1,2 \ldots, u,$$

$$Sum \sum_{k=1}^{u} X_{kl} = q_l \qquad l = 1,2 \ldots, v,$$

$$X_{kl} \geq 0 \quad k = 1, 2 \ldots, u; l = 1,2 \ldots, v.$$

7.3 IMPROVED ALGORITHM

Step 1: Use the previously mentioned center of gravity method to convert all trapezoidal fuzzy data from the given fuzzy transportation table into their equivalent crisp values.

Step 2: Verify whether the given fuzzy transportation problem is balanced. If it is balanced, proceed to Step 3. If it is not balanced, balance it by adding a suitable dummy row or column, and then proceed to Step 3.

Step 3: Subtract the minimum value from the maximum value in each row. Do the same for each column.

From the resulting values, choose the highest value. Then, select the minimum value in the corresponding row or column.

Step 4: Determine the minimum value between the supply and demand for the selected cell and assign it accordingly. Continue applying Step 4 until all supply and demand constraints are satisfied.

Step 5: To check the optimality of the transportation problem, solve using the formula $c_{ij} = u_i + v_j$, and determine u_i and v_j for each occupied cell (i,j). Determine d_{ij}, where $d_{ij} = c_{ij} - (u_i + v_j)$, for each vacant cell (i,j). As all $d_{ij} \geq 0$. Thus, the ultimate minimization value has been found.

Time complexity: The time complexity of the improved algorithm for solving the fuzzy transportation problem is $O(m \times n)$, where m is the number of sources and n is the number of destinations. The algorithm involves converting fuzzy data to crisp values, checking for balance, performing allocations based on cost comparisons, and verifying optimality using dual variables and opportunity costs. Since each step requires scanning or processing each cell of the transportation table at most a constant number of times, the dominant operations are proportional to the number of cells, leading to an overall time complexity of $O(m \times n)$.

7.4 NUMERICAL EXAMPLE

These examples (Tables 7.2 and Table 7.10) of a transportation problem illustrate how to solve a fuzzy transportation problem where transportation costs, supply, and demands are represented by trapezoidal fuzzy numbers. The cost matrix between the transportation problem's supply sites and demand destinations is shown in Table 7.1, and Figure 7.2 shows the transportation model's structure and every conceivable path between sources and destinations.

Table 7.1 Transportation Problem of Three Sources and Four Demands.

	T1	T2	T3	T4	*Demand*
S1	D_{11}	D_{12}	D_{13}	D_{14}	
S2	D_{21}	D_{22}	D_{23}	D_{24}	
S3	D_{31}	D_{32}	D_{33}	D_{34}	
Supply					

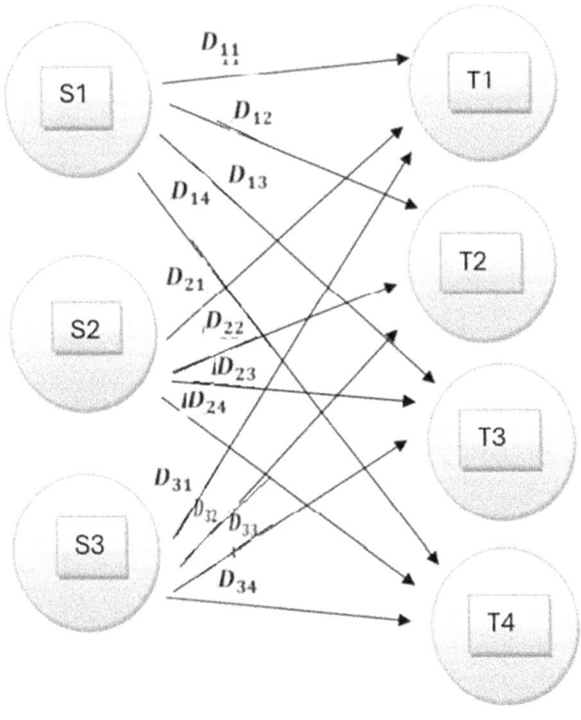

Figure 7.2 Transportation Problem.

7.4.1 Numerical Example 1

A fuzzy transportation problem with three sources and four destinations is discussed in this example. Table 7.2 shows the triangular fuzzy expenses associated with each route.

Using the chosen defuzzification technique, the triangular fuzzy cost values shown in Table 7.2 are transformed into their respective crisp equivalents, which are then shown in Table 7.3.

Following the implementation of Steps 2–4 of the process, which include supply–demand updates, cell selection, and index computation, Table 7.4 displays the final allocation.

$$\text{Min } z = 8 + 68 + 21 + 63 + 66 + 5 = 231$$

The allocation produced by using the least cost method is shown in Table 7.5.

$$\text{Min } z = 27 + 51 + 18 + 49 + 176 + 42 = 363$$

Table 7.2 Trapezoidal Fuzzy Numbers.

	T1	T2	T3	T4	*Demand*
S1	(1,3,5,7)	(1,3,5,7)	(8,10,12,14)	(11,13,15,17)	(16,18,20,22)
S2	(0,2,4,6)	(3,5,7,9)	(5,7,9,11)	(4,6,8,10)	(13,15,17,19)
S3	(5,7,9,11)	(0,2,4,6)	(0,2,4,6)	(2,4,6,8)	(20,22,24,26)
Supply	(6,8,10,12)	(14,16,18,20)	(17,19,21,23)	(7,9,11,13)	

Table 7.3 Transformation of Trapezoidal Fuzzy Numbers to Crisp Numbers

	T1	T2	T3	T4	*Demand*
S1	4	4	11	14	19
S2	3	6	8	7	16
S3	8	3	3	5	23
Supply	9	17	22	10	

Table 7.4 Final Allocation.

	T1	T2	T3	T4	*Demand*
S1	4 2	4 17	11	14	0
S2	3 7	6	8	7 9	0
S3	8	3	3 22	5 1	0
Supply	0	0	0	0	

Table 7.5 Final Allocation Using the Least Cost Method.

	T1	T2	T3	T4	*Demand*
S1	4	4	11	14	0
			16	3	
S2	3	6	8	7	0
	9			7	
S3	8	3	3	5	0
		17	6		
Supply	0	0	0	0	

Table 7.6 Final Allocation Using the Northwest Corner Method.

	T1	T2	T3	T4	*Demand*
S1	4	4	11	14	0
	9	10			
S2	3	6	8	7	0
		7	9		
S3	8	3	3	5	0
			13	10	
Supply	0	0	0	0	

Table 7.7 Final Allocation Using the Southwest Corner Method.

	T1	T2	T3	T4	*Demand*
S1	4	4	11	14	0
			9	10	
S2	3	6	8	7	0
		3	13		
S3	8	3	3	5	0
	9	14			
Supply	0	0	0	0	

The allocation produced by using the Northwest corner method is shown in Table 7.6.

$$\text{Min } z = 36 + 40 + 42 + 72 + 39 + 50 = 279$$

The allocation produced by using the Southwest corner method is shown in Table 7.7.

$$\text{Min } z = 140 + 99 + 104 + 18 + 42 + 72 = 475$$

The allocation produced by using Vogel's approximation method (VAM) is shown in Table 7.8.

$$\text{Min } z = 68 + 27 + 66 + 28 + 49 + 5 = 243$$

Table 7.8 Final Allocation Using VAM.

	T1	T2	T3	T4	*Demand*
S1	4	4 17	11	14 2	0
S2	3 9	6	8	7 7	0
S3	8	3	3 22	5 1	0
Supply	0	0	0	0	

Table 7.9 Optimal Table.

	T1	T2	T3	T4	*Demand*
S1	4	4 17	11	14 2	0
S2	3 9	6	8	7 7	0
S3	8	3	3 22	5 1	0
Supply	0	0	0	0	

Step 5 computes the optimal value of the specified transportation problem, and Table 7.9 presents the optimal table that results.

$$\text{Min } z = 8 + 68 + 21 + 63 + 66 + 5 = 231$$

The result compared using LINGO software WIN32 20.0.29

Figure 7.3 displays the optimal output produced by LINGO for Example 7.4.1.

Min z = 231

7.4.2 Numerical Example 2

A fuzzy transportation problem with three sources and four destinations is discussed in this example. Table 7.10 shows the triangular fuzzy expenses associated with each route.

Using the chosen defuzzification technique, the triangular fuzzy cost values shown in Table 7.10 are transformed into their respective crisp equivalents, which are then shown in Table 7.11.

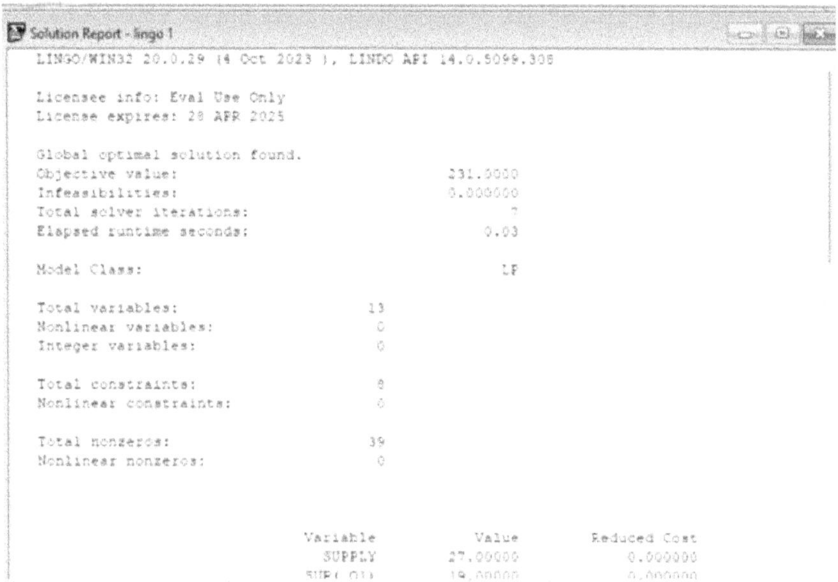

Figure 7.3 Obtained the optimal value using LINGO software for Example 7.4.1.

Table 7.10 Trapezoidal Fuzzy Numbers.

	T1	T2	T3	T4	*Demand*
S1	(2,4,6,8)	(2,4,6,8)	(9,11,13,15)	(12,14,16,18)	(17,19,21,23)
S2	(1,3,5,7)	(4,6,8,10)	(6,8,10,12)	(5,7,9,11)	(14,16,18,20)
S3	(6,8,10,12)	(1,3,5,7)	(1,3,5,7)	(3,5,7,9)	(21,23,25,27)
Supply	(7,9,11,13	(15,17,19,21)	(18,20,22,24)	(8,10,12,14	

Table 7.11 Trapezoidal Fuzzy Numbers to Crisp Numbers.

	T1	T2	T3	T4	*Demand*
S1	5	5	12	15	20
S2	4	7	9	8	17
S3	9	4	4	6	24
Supply	10	18	23	11	

Following the implementation of Steps 2–4 of the process, which include supply–demand updates, cell selection, and index computation, Table 7.12 displays the final allocation.

$$\text{Min } z = 10 + 90 + 32 + 72 + 92 + 6 + 0 = 302$$

The allocation produced by using the least cost method is shown in Table 7.13.

$$\text{Min } z = 40 + 85 + 4 + 92 + 45 + 56 + 0 = 322$$

The allocation produced by using the Northwest corner method is shown in Table 7.14.

$$\text{Min } z = 50 + 50 + 56 + 81 + 56 + 60 = 353$$

The allocation produced by using the Southwest corner method is shown in Table 7.15.

$$\text{Min } z = 165 + 108 + 126 + 21 + 81 + 60 = 561$$

Table 7.12 Solution by Using the Given Algorithm Steps 2–4.

	T1	T2	T3	T4	*Demand*
S1	5 2	5 18	12	15	0
S2	4 8	7	9	8 9	0
S3	9	4	4 23	6 1	0
S_{dummy}	0	0	0	0 1	0
Supply	0	0	0	0	

Table 7.13 Solution by Using the Least Cost Method.

	T1	T2	T3	T4	*Demand*
S1	5	5 17	12	15 3	0
S2	4 10	7	9	8 7	0
S3	9	4 1	4 23	6	0
S_{dummy}	0	0	0	0 1	0
Supply	0	0	0	0	

Table 7.14 Solution by Using the Northwest Corner Method.

	T1	T2	T3	T4	*Demand*
S1	5	5	12	15	20
	10	10			
S2	4	7	9	8	17
		8	9		
S3	9	4	4	6	24
			14	10	
S_{dummy}	0	0	0	0	0
				1	
Supply	10	18	23	11	20

Table 7.15 Solution by Using the Southwest Corner Method.

	T1	T2	T3	T4	*Demand*
S1	5	5	12	15	20
			9	11	
S2	4	7	9	8	17
		3	14		
S3	9	4	4	6	24
	9	15			
S_{dummy}	0	0	0	0	0
	1				
Supply	10	18	23	11	

Table 7.16 Solution by Using VAM.

	T1	T2	T3	T4	*Demand*
S1	5	5	12	15	0
		18		2	
S2	4	7	9	8	0
	10			7	
S3	9	4	4	6	0
			23	1	
S_{dummy}	0	0	0	0	0
				1	
Supply	0	0	0	0	

The allocation produced by using VAM is shown in Table 7.16.

$$\text{Min } z = 90 + 30 + 40 + 56 + 92 + 6 = 314$$

Step 5 computes the optimal value of the specified transportation problem, and Table 7.17 presents the optimal table that results.

$$\text{Min } z = 10 + 90 + 32 + 72 + 92 + 6 + 0 = 302$$

Table 7.17 Optimal Table.

	T1	T2	T3	T4	*Demand*
S1	5	5	12	15	0
	2	18			
S2	4	7	9	8	0
	8			9	
S3	9	4	4	6	0
			23	1	
S_{dummy}	0	0	0	0	0
				1	
Supply	0	0	0	0	

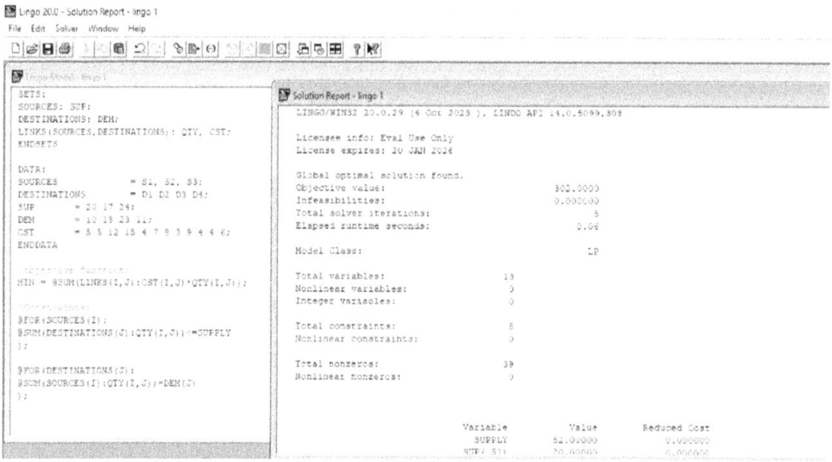

Figure 7.4 Obtained the optimal value using LINGO software, for Example 7.4.2.

The result compared using LINGO software WIN32 20.0.29.

Figure 7.4 displays the optimal output produced by LINGO for Example 7.4.2.

$$\text{Min } z = 302$$

7.5 COMPARATIVE SURVEY

By converting the given problems into clear equivalents, our proposed method shows that it is effective in solving fuzzy transportation problems and lowers transportation expenses when compared to certain other existing methods, such as the least cost method, the Northwest corner method, the Southwest corner method, and VAM. We also solved these numerical examples using LINGO software, which yielded the same results as our proposed algorithm.

The suggested strategy is compared with other strategies in Tables 7.18 and 7.19, and Figures 7.5 and 7.6 provide a visual representation of the same comparison.

Table 7.18 Comparative Table for Example 7.4.1.

Method	*cost*
Least cost method	363
Northeast corner method	279
Southwest corner method	475
VAM	243
Proposed method	231
Optimal value	231
Optimal value using LINGO software	231

Table 7.19 Comparison Table for Example 7.4.2.

Method	*Cost*
Least cost method	322
Northeast corner method	353
Southwest corner method	561
VAM	314
Proposed method	302
Optimal value	302
Optimal value using LINGO software	302

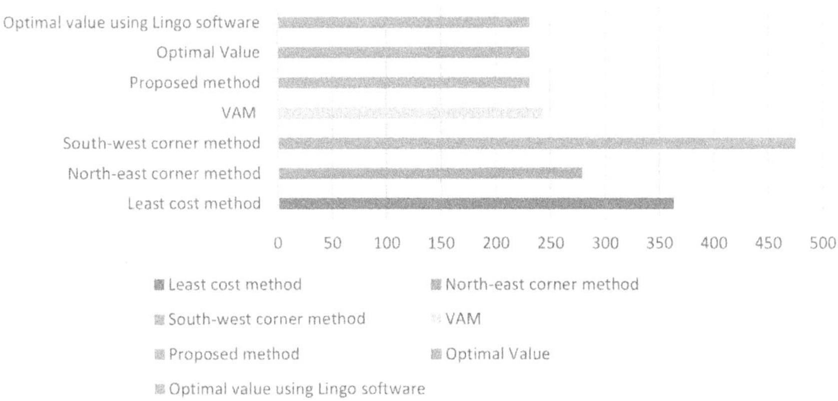

Figure 7.5 Comparison graph of Example 7.4.1.

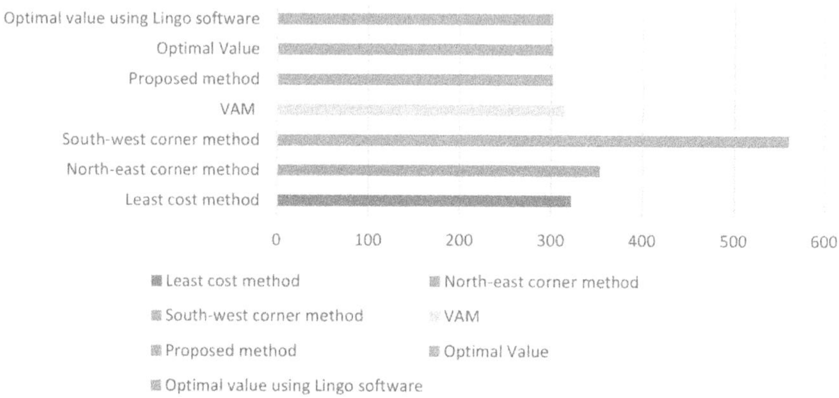

Figure 7.6 Comparison graph of Example 7.4.2.

7.6 CONCLUSION

This chapter presents a solution to transportation problems in a trapezoidal fuzzy environment. By converting the fuzzy problem into a crisp equivalent, we obtained the optimal solution using our proposed method. A comparative analysis with existing methods highlights the novelty and effectiveness of our approach. A simple yet very efficient technique is presented in this chapter to tackle the fuzzy transportation problem. Trapezoidal fuzzy numbers are explained with appropriate examples. The approach proposed in this chapter applies to all types of fuzzy transportation problems. Moreover, with additional fuzzy algorithms, this approach can be expanded. Fuzzy Trapezoidal Representation: We are introducing fuzzy trapezoidal numbers to represent uncertain transportation costs, demand, and supply. This creates a more realistic and robust model that reflects the complexities of real-world transportation systems. Hybrid Algorithm: Our research creates a hybrid algorithm that combines fuzzy arithmetic operations with LP techniques. This potent approach guarantees efficient and effective solutions to transportation problems. For those who must make decisions in unpredictable situations, this offers insightful information. To show the applicability and efficacy of our suggested method in managing uncertainty, it is applied to a real-world transportation situation. Future Research Directions: Potential avenues for future research include extension to other fuzzy numbers, i.e., investigating the application of other fuzzy numbers, such as fuzzy triangular or fuzzy Gaussian numbers, to transportation problems; multi-objective optimization, i.e., developing a multi-objective optimization framework to address competing objectives in transportation problems, such as cost, time, and environmental impact; and real-time applications, i.e., exploring the application of fuzzy trapezoidal numbers in real-time transportation systems, such as traffic management and logistics.

Declaration of Competing Interest: The authors affirm that none of the work described in this publication may have been influenced by any known competing financial interests or personal relationships.

REFERENCES

1. Bellman, R. E., & Zadeh, L. A. (1970). Decision-making in a fuzzy environment. *Management Science*, *17*(4), B–141.
2. Hillier, F. S., & Lieberman, G. J. (2005). "Non-linear programming" and "metaheuristics". In *Introduction to Operations Research* (Vol. 13, pp. 547–616). McGraw-Hill Publication.
3. Zadeh, L. A. (1965). Fuzzy sets. *Information and Control*, *8*(3), 338–353.
4. Gani, A. N., & Razak, K. A. (2004). Solving transportation problem using fuzzy number. *Bulletin of Pure and Applied Science E*, *23*, 281–289.
5. Zimmermann, H. J. (1978). Fuzzy programming and linear programming with several objective functions. *Fuzzy Sets and Systems*, *1*(1), 45–55.
6. Fang, S. C., Hu, C. F., Wang, H. F., & Wu, S. Y. (1999). Linear programming with fuzzy coefficients in constraints. *Computers & Mathematics with Applications*, *37*(10), 63–76.
7. Parra, M. A., Terol, A. B., & Urıa, M. R. (1999). Solving the multiobjective possibilistic linear programming problem. *European Journal of Operational Research*, *117*(1), 175–182.
8. Luhandjula, M. K. (1987). Linear programming with a possibilistic objective function. *European Journal of Operational Research*, *31*(1), 110–117.
9. Behera, S. P., Pati, J. K., Raut, P. K., & Mahanta, K. L. (2022). Calculation of linear fractional fuzzy transportation problem using simplex method. *International Journal of Engineering Science Technologies*, *6*(2), 38–47.
10. Bharati, S. K. (2021). Transportation problem with interval-valued intuitionistic fuzzy sets: impact of a new ranking. *Progress in Artificial Intelligence*, *10*(2), 129–145.
11. Muruganandam, S., & Srinivasan, R. (2016). A new approach for solving unbalanced fuzzy transportation problem. *Asian Journal of Research in Social Sciences and Humanities*, *6*(5), 673–680.
12. Srinivasan, B., Ganeswaran, D. P., & Meenambal, D. T. (2016). Optimization with sun light source in old constructed building and converting to green building. *International Journal of Civil Engineering and Technology*, *7*(5), 428–434.
13. Więckowski, J., Kizielewicz, B., Wątróbski, J., & Sałabun, W. (2024). A new approach for handling uncertainty of expert judgments in complex decision problems. *IEEE Access*, *12*, 142026–142046.
14. Pooja, M., Thangapandi, C., & Srinivasan, R. (2024, May). New algorithmic approach for solving transportation problem in fuzzy environment with trapezoidal fuzzy numbers. In *American Institute of Physics Conference Series* (Vol. 2853, No. 1, p. 020057).
15. Srinivasan, R., Karthikeyan, N., & Jayaraja, A. (2021). A proposed technique to resolve transportation problem by trapezoidal fuzzy numbers. *Indian Journal of Science and Technology*, *14*(20), 1642–1646.

16. Dhouib, S. (2021). Solving the trapezoidal fuzzy transportation problems via new heuristic: The Dhouib-Matrix-TP1. *International Journal of Operations Research and Information Systems (IJORIS)*, *12*(4), 1–16.
17. Bejines, C., & Navara, M. (2022). The Fibonacci sequence in the description of maximal discrete Archimedean t-norms. *Fuzzy Sets and Systems*, *451*, 94–112.
18. Ghosh, S., Roy, S. K., Ebrahimnejad, A., & Verdegay, J. L. (2021). Multi-objective fully intuitionistic fuzzy fixed-charge solid transportation problem. *Complex & Intelligent Systems*, *7*(2), 1009–1023.
19. Akram, M., Shah, S. M. U., Al-Shamiri, M. M. A., & Edalatpanah, S. A. (2022). Fractional transportation problem under interval-valued Fermatean fuzzy sets. *AIMS Math*, *7*(9), 17327–17348.
20. Baranidharan, B., & Mahapatra, G. S. (2024). A new solution technique for fuzzy transportation problem using novel ranking functions on heptagonal fuzzy numbers: A case study of regional shipment. *Journal of Computational and Cognitive Engineering*.

Chapter 8

Analyzing the Selection Parameters to Establish a Choice of Retail Store (Hypermarket, Convenience Store, or Supermarket) in a Target Location

A Fuzzy-AHP Approach

Amit Yadav and Rahul Priyadarshi

8.1 INTRODUCTION

Brand-owned retail stores are established in a locality not only to earn profit but also to provide the populace with standard products at competitive pricing and to offer employment opportunities [1–3]. A brand always decides on the scale of the retail outlet, i.e., a supermarket, a hypermarket, or a convenience store, to be established based on real estate pricing, acquisition, leasing, or monthly incurred costs, considering potential revenues earned from the specific geography [1, 4, 5]. A supermarket holds the capacity to store more non-perishables with a strategy of everyday low pricing (EDLP). It has been observed that a supermarket typically acquires real estate for a longer duration to reduce the costs incurred over time. A hypermarket is established to serve the public with a wider selection of food items in a single location, thereby reducing the need for commuting to buy groceries. Hypermarkets may or may not own the real estate, depending on their strategic decisions regarding costs, premium pricing, and expansion projections. Convenience stores today offer a mix of food and nonfood items that consumers may require on a day-to-day basis, eliminating the need to travel longer distances to purchase supplies. These convenience stores are typically leased and carry a limited inventory of consumer goods with maximum retail prices (MRP). It then falls upon consumers to decide whether to patronize these convenience stores or not, based on their buying behavior and offered pricing. If the retail stores have made a miscalculation of the consumption pattern, target consumer attributes, population density, customer needs, or consumers' purchasing capability, the product placement cannot translate into sales [4, 6]. Consumers today expect a home delivery of fast-moving consumer goods (FMCG) with competitive commodity and delivery charges. Hence, all these retail formats have initiated efforts to practice omnichannel

DOI: 10.1201/9781003642091-8

operations, i.e., to operate both physical retail stores and online delivery options, to serve their customers from these retail stores. However, not every retail outlet has understood consumer demand, patterns, or pricing strategies appropriately [4, 5]. Hence, there are chances that they ignore the subtle signals of low demands and leftovers till they realize that their business is going through a deficit due to inefficient decision-making. Hence, costs and market demand, footfall, and purchasing capability should be assessed well in advance. One of the reasons these retail outlets face losses is also due to handling inaccurate amounts of inventory, ignoring seasonal influxes, and subsequent demand trends. Hence, a specific amount of inventory shall be ordered and procured by these retailers so that there are no leftovers at the end of the desired period, thereby avoiding unnecessary costs [3, 6]. The firms can also practice definitive milk runs for effective distribution [1, 2, 7, 8]. Operational and tactical level decisions, such as product placement, inventory flow, and seasonal/nonseasonal offers for a brand-owned retail outlet, are usually made at the regional headquarters. Meanwhile, the headquarters can also fail to accurately assess the consumers' purchasing capability. Hence, these losses occur. Overall, decision-makers should also consider product spoilage and waste to ensure a firm achieves its stated sustainability goals [9].

8.1.1 Research Question

The decision to establish a store also depends on the market presence of competitors. This research examines the current presence, advantages, disadvantages, and market share of retail businesses. The study highlights opportunities for a new retail model to stand out by identifying market gaps and evaluating competitive pricing, product mix, and service quality. Customers experience numerous retail experiences, such as self-service, home delivery, or loyalty programs, that enable them to pick a model that delivers the best competitive advantage [10].

The perfect functioning of a retail organization depends on the capability of its supply network. Local and regional suppliers' dependability and availability enable one to assess the viability of inventory replenishment and stocking fairly. Logistically, every type of retail is evaluated for its infrastructure, including warehouses, distribution centers, and transportation systems, to identify any benefits or drawbacks [11]. Every structure—including real estate expenses, store layout fees, and product acquisition—should be assessed generally in terms of capital expenditure. Analyzes of operational costs, including rent, labor, utilities, and supply chains, enables one to forecast the ongoing financial load. Break-even analysis, profitability margins, and income possibilities are included in every retail layout to forecast the expected return on investment (ROI). This financial analysis allows retailers and investors to find Pune's most reasonably priced and environmentally friendly retail strategy.

A few research questions are framed for the study:

RQ1: Do retail chains operate based on the market demand of the geographical area?
RQ2: Do retail chains perform the overall market need assessment before establishing a store?
RQ3: Do retail chains take cognizance of consumer price sensitivity and spending habits before establishing a store?

This study attempts to address the above research questions.

8.2 LITERATURE REVIEW

Many research studies have examined the effect of architectural surroundings on food purchases. This study motivates the real estate sector/business strategy researchers to become more active by offering a thorough body of information on retail store placement and market analysis. Defining the region involves macro-level market selection strategies, micro-simulating consumer behavior, and consumer behavior in models of shop placement. Recent research on retail leasing and space allocation in shopping complexes shows the progress of work on retail store placement and market analysis [1]. The strategic decision-making process of choosing a retail company location was investigated using research and a long-term investment impacting consumer happiness and profitability. It separates stores based on suggestions and looks at historical research. Using the analytic hierarchy process (AHP), a selection model was developed in real-world business life to identify the most suitable site for retail stores using qualitative and quantitative techniques [4]. The effect of rental value and occupancy expenditures on profitability was sought in the research to provide independent enterprises and shopping malls with a whole profitability calculation. The study used a simple profitability model and secondary data from one retail organization. It looks at the store size and rental value to assess the profitability of numerous shop ideas. The research emphasizes, for best outcomes, the need to combine consumer needs with financial benchmarking data [3]. The Delphi method was adapted to evaluate marketing strategies and areas of franchise management for convenience stores. A total of 120 franchisees provided Delphi responses during the interview. The franchises were performing marginally due to a lack of inventory and the need for continuous sales promotion activity. They observed unpleasant shop visits and ineffective problem-solving from consumers [8]. A study covering 300 Chennai-based consumers indicated that hypermarkets are growing in popularity for their variety and branded products. In all, 65% of consumers are comfort seekers, and 35% are product searchers. The study revealed that hypermarkets attract consumers more than supermarkets because of their wide range of items and additional services than their pricing, which might motivate impulsive buying behavior [2]. Racial and poverty characteristics influence food store closures in US cities; high-poverty shops have shorter living lives, and non-White areas experience disadvantages. Consequently, there may be a confusing understanding of the association between shop availability and health outcomes [6]. Analyzing US family food consumption found no relationship between store accessibility and fruit and vegetable consumption. Additionally, fruit purchase was negatively correlated with convenience stores ($p = 0.001$) [6]. Environmental factors, road connectivity, and regional destination accessibility impact purchase behavior. Policies meant to boost supermarket availability might not be enough to inspire fruit and vegetable intake. This study showed that healthier eating habits need a more all-encompassing approach involving site variation and accessibility [12]. A study examined customer impressions of retail firm locations and their impact on service quality. It questioned 1,055 consumers in three Amman superstores: Carrefour, Safeway, and CTown. According to the study, retail site size has little impact on servicescape quality. The results highlight that experience at the store surpasses word-of-mouth publicity [5]. Kwa Mashu looked at variations in the buying patterns of the consumers of an independent convenience store based on the design of the business. A quantitative study on self-administered surveys reveals that customers struggle with appearance, product presentation, retail environment, in-store service, and shop layout accessibility. The findings suggest that the management of Kwa Mashu convenience shops should focus on these areas and use procedures to support customers as necessary to boost turnover, customer loyalty, and sales [7]. Real estate experts are motivated to become more involved in macro-level market selection tactics, micro-simulating consumer behavior, consumer behavior, and company placement as this assists in decision-making on retail locations and market analysis [9]. The study also offered a fresh method for selecting local community convenience store locations aligned with sustainability and unpredictable behavior. The enhanced gray wolf optimization method increases convergence, stability, and search accuracy. Testing on Suzhou Street in Xiaonanzhuang district of Beijing, the model integrates macroenvironment, income, and cost to exhibit its scientific, practicality, and sustainability aspects [9].

8.2.1 Objective

This study uses a fuzzy analytic hierarchy process (FAHP)-based decision-making technique to determine Pune's most appropriate retail model. Under this rigorous approach, which is grounded in many criteria, the study evaluates selection over hypermarkets, supermarkets, and convenience shops in search of the best retail model. AHP has been used by similar studies, such as [4] for location selection in Turkey based on election/population data. [13] used FAHP to decide the level of vertical integration in agribusiness at the rural level. The research assesses major customer preferences and buying behavior to better understand purchase patterns, number of visits, and product preferences that influence the choice of store form. Moreover, the study evaluates the long-term viability of retail companies considering operating expenses, population density, market demand, and legal restrictions in Pune [14]. This study assists in deciding if more minor, more local businesses would be appropriate or whether large retail stores hold the potential to grow in the area. Ultimately, the research provides a direction to stores looking to expand data-driven advice, thus assuring that their business selections suit customer wants, financial situations, and legal requirements. This thorough investigation will comprehensively help businesses achieve a strong retail presence in Pune.

8.3 PROPOSED METHODOLOGY

The proposed methodology involves tracking the experts' perspectives based on present infrastructure, government support, and expanding innovations trends using the FAHP method. The research followed the approach outlined in Figure 8.1.

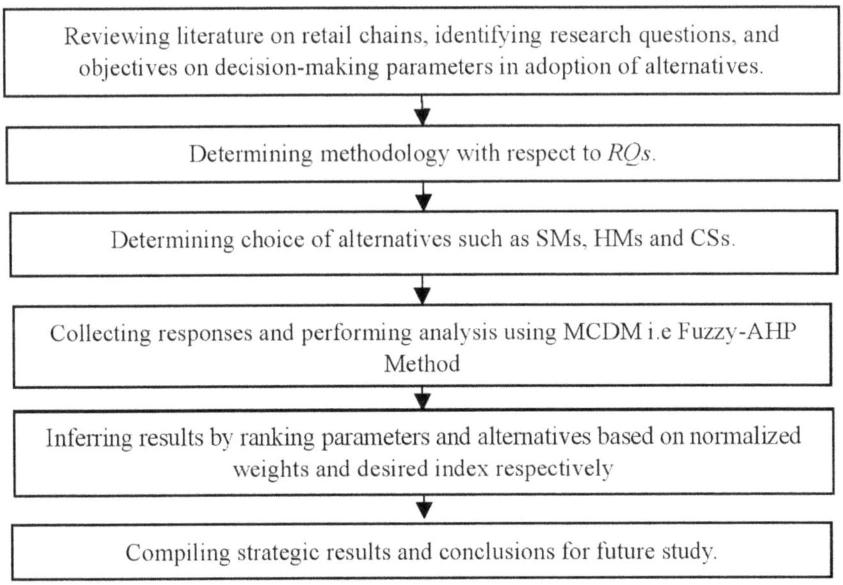

Figure 8.1 Flowchart to evaluate various retail configurations over parameters.

The following equation is used to determine the correctness of the information.

$$\text{Sample} = \left(z^2\right) \times \left[\frac{p \times q}{e^2}\right] \tag{8.1}$$

Here, $z = 1.645$ (= 90% confidence); $p =$ population familiar with the issue, i.e., 0.9; $q = (p - 1) = 0.1$, i.e., population unfamiliar with the problem; and *error*, $e = 10\%$. Here, z = normalization value derived from the z-table. Consequently, with a small sample size, the final result shows 24.35. The number of expert responses considered for the pairwise comparison is highlighted in Table 8.1.

The literature review for parameters is presented in Table 8.2.

Table 8.1 Experts' Survey Diversity

Sr. No.	*Profession*	*Education*	*Experience (Years)*	*Frequency*
1.	Senior supply chain managers	Graduation + postgraduation	10–15	5
2.	Retail managers	Graduation + postgraduation	10–15	8
3.	Warehouse executives	Graduation	5–10	6
4.	Store-level analysts	Graduation + postgraduation	5–10	5
5.	Supply chain scientists	Postgraduation + PhD	5–10	6
			Total	**30**

Table 8.2 Literature Review for Parameters.

Sr. No.	*Factor/Parameter*	*Notation*	*Literature Review*
1	Consumption pattern	CPR	[5, 7, 9, 15]
2	Target consumer attributes	TCA	[3, 5, 7]
3	Population density	PDP	[6, 12]
4	Foot traffic and accessibility	FCA	[3, 5, 7, 12]
5	Competitive supply network	CSN	[1, 8]
6	Evolving customer needs	ECN	[2, 3, 8, 13]
7	Shelf space allocation	SSA	[1, 3, 5]
8	Fulfilment rate	FRR	[2, 7]
9	Supply chain overhead	SCO	[3, 9, 13]
10	Consumer's purchasing capability	CPC	[7, 12]
11	Compliance framework	CPF	[8, 9, 12]

Analytical Hierarchy Process Steps

A systematic decision-making tool called the analytic hierarchy process (AHP) enables one to assess many possibilities depending on the criteria. The procedure consists of the following main stages:

1. *Define the problem and set the hierarchy in the first step*: Clearly defining the choice problem—i.e., choosing the most appropriate retail format—hypermarket, supermarket, or convenience store—for Pune—comes first. Then, a hierarchical framework is developed with the finding of the ideal shop layout as the level goal. Key elements impacting retail choice, including market demand, competitiveness, supply chain capabilities, regulatory limits, and financial feasibility, define an intermediary level. Three retail models—hypermarket, supermarket, and convenience store—are being examined from bottom-level alternatives.
2. *Perform pairwise comparisons with input information:* The pairwise comparisons are performed with surveys conducted with experts, business leaders, and stakeholders. Saaty's 9-point scale is utilized, where

   1 = equal importance
   3 = moderate importance
   5 = strong importance
   7 = very strong importance
   9 = extreme importance.

   The intermediate values (2, 4, 6, 8) are used for finer distinctions.
3. *Creating pairwise comparison matrices:* Pairwise comparison matrices are developed for every criterion at many hierarchy tiers based on the expert answers. These matrices enable the measurement of decision-maker preferences by reflecting the relative weights given to every constituent.
4. *Calculate the priority weights*: Regarding every pairwise comparison matrix, divide every value in the matrix by the column total to normalize it. To obtain the relative importance—i.e., the priority weight—of every criterion and option, average each row. This produces a set of numerical weights reflecting the relative relevance of alternatives' ranking and the criteria's significance.

   The triangular fuzzy approach was used as it assists in explaining the inputs to variables more precisely, as the response to input voids is minimized, and also due to its ease of use, efficiency, and ability to capture the defined peak point. Figure 8.2 highlights the same.

   Here, $f_n(x)$ is the membership function; k is the minimum fuzzy number; m is the maximum fuzzy number; and l is the mean fuzzy estimation.

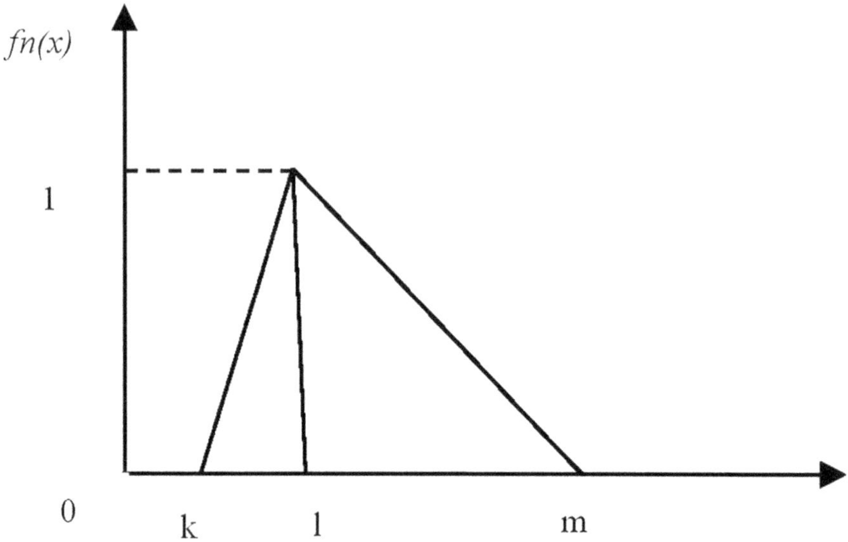

Figure 8.2 Representation of the efficiency and ability to capture the peák point.

5. *Seeking consistency:* Reliable decision-making depends on consistency in pairwise comparisons; the consistency index (CI) is calculated using the following equation.

$$CI = \frac{(\lambda_{max} - n)}{(n-1)}, \tag{8.2}$$

where n is the number of criteria or options, and λ_{max} is the greatest eigenvalue of the comparison matrix.

The consistency ratio (CR) is thus computed from

$$CR = \frac{CI}{RI}. \tag{8.3}$$

A standard value derived from the matrix size is the random index (RI), and CR should be less than 0.10; the comparisons align. If not, the comparisons are looked over and corrected.

6. *Figuring out final scores and rank alternatives:* The final scores for every retail form are calculated by aggregating the priority weights from many hierarchy tiers. The most appropriate retail arrangement for Pune is the one with the best total score.

Following these methodical guidelines, AHP offers Pune a methodical, objective, and data-driven choice for the best retail layout.

8.4 RESULTS AND DISCUSSION

The study examines the identified selection parameters that are helpful for strategic decision-making to establish supermarkets, hypermarkets, and convenience stores. This qualitative analysis was performed by collecting information from an array of experts/stakeholders for their opinions. Then the steps mentioned above were followed to draw results. Under certain strategic paradigms, the calculated integrated weights help one grasp every piece's relative importance, as highlighted in Table 8.3. A donut diagram is also provided in Figure 8.3 to highlight the level of preference to choose convenience stores over supermarkets and hypermarkets.

8.4.1 Analysis of Individual Parameters

- Foot traffic and accessibility (FCA) has obtained the highest NW score of 0.304, which implies that the volume of consumer footfall to the store and conversion into sales is one of the prime elements.
- Target consumer attributes (TCA) achieve the second-level priority score of 0.136, as adapting and serving as per changing consumer trends are very important.
- Consumption pattern (CPR) has scored the third priority score of 0.113 on the scale, as capturing demand patterns and consumer behavior is crucial.
- Population density (PDP) is also one of the key factors to be considered, as it highlights the potential of business with respect to the proximity of the store for all the retail configurations and people visiting them, with a score of 0.097.
- Competitive supply network (CSN) and Fulfilment Rate (FRR) are the operational parameters that a supply chain needs to ensure when all the above factors drive the sales trend, with scores 0.089 and 0.083. Hence, making products available as per the supply chain strategy is one of the necessities.

8.4.2 Parameter-Wise Insights

- FCA is the most important factor to establish a convenience store, a supermarket, and a hypermarket, with scores of 0.292, 0.245, and 0.128, respectively.
- CPR is a very crucial metric to be considered to establish a convenience store, a supermarket, and a hypermarket, with scores of 0.169, 0.118, and 0.047, respectively.
- The rest of the factors also hold their crucial value that are colloquial with the above metrics. Hence, it shall be focused upon.

Table 8.3 Normalized Weights of Alternatives and Parameters.

Parameter	*Normalized Weights of Parameter*	*Normalized Weights against Each Parameter*			*Integrated Weights*		
		Supermarkets	*Hypermarkets*	*Convenience Stores*	*Supermarkets*	*Hypermarkets*	*Convenience Stores*
Consumption pattern (CPR)	0.113	0.118	0.047	0.169	0.013	0.005	0.019
Target consumer attributes (TCA)	0.136	0.113	0.090	0.000	0.015	0.012	0.000
Population density (PDP)	0.097	0.111	0.088	0.081	0.011	0.009	0.008
Foot traffic and accessibility (FCA)	0.304	0.245	0.128	0.292	0.075	0.039	0.089
Competitive supply network (CSN)	0.089	0.103	0.086	0.071	0.009	0.008	0.006
Evolving customer needs (ECN)	0.066	0.097	0.084	0	0.006	0.006	0
Shelf space allocation (SSA)	0.055	0.027	0.042	0.165	0.002	0.002	0.009
Fulfilment rate (FRR)	0.083	0.016	0.082	0.161	0.001	0.007	0.013
Supply chain overhead (SCO)	0.044	0.005	0.080	0.000	0	0.004	0
Consumer's purchasing capability (CPC)	0.012	0.091	0.077	0.000	0.001	0.001	0
Compliance framework (CPF)	0.001	0.074	0.074	0	0	0	0
Desired index (DI 1,2,3)					**0.13**	**0.09**	**0.14**

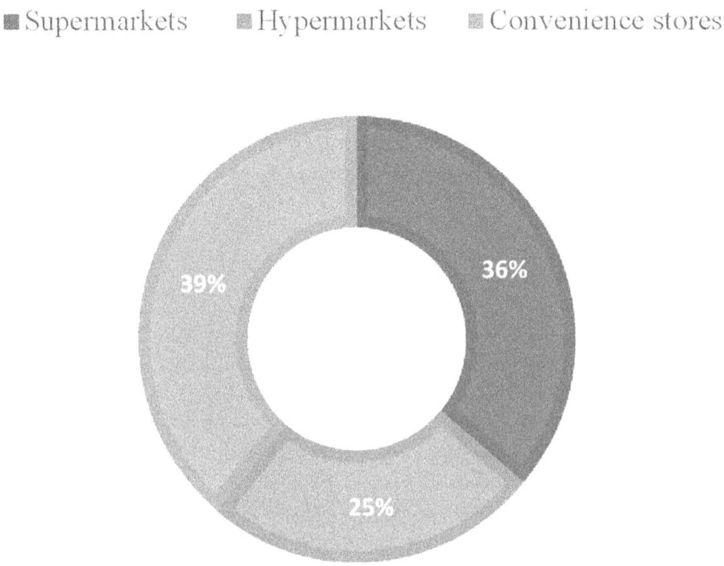

Figure 8.3 A donut chart of DI values for level prioritization.

8.4.3 Strategic Results with Combined Weights

The DI values highlight that, in the given conditions, establishing a convenience store holds the maximum priority since it requires less economic capability but shall be established as per the needs of the consumers. The supermarkets hold the second priority as they require an economic capital decision for a long-term investment. Consumer footfall, adapting to consumer attributes and CPRs, is also crucial.

8.5 CONCLUSION

The retail environment today faces significant competition, not only from the offline supply chain but also from manufacturer-owned direct delivery channels. Hence, it is crucial to observe consumer sentiment, demand, and proximity. In the era of rapidly fluctuating demand trends, retailers and businesses must closely monitor daily and weekly demand, as well as their competitors' product placement. The concept of omnichannel retailing is the need of the hour; one needs to practice it in conjunction with strategic offline retailing to serve the consumer segment. Thus, the above analysis helps focus on the given parameters in a prioritized manner. Eventually, with better insights, better operational planning, and improved consumer focus, the retail outlets of various shapes will be able to sustain themselves. There are examples of supermarkets that are highly successful despite cut-throat competition from online retail and quick-commerce ventures, which offer lower prices and wider product offerings. However, a few hypermarkets and supermarkets had to close down due to inefficient operations, pricing, and consumer representation. Thus, organized retail ventures need to intervene in strategic decision-making to stay strong in the market with price-effective home-grown brands and omnichannel delivery.

REFERENCES

1. K. Vandall, and C. Carter, "Retail Store Location and Market Analysis: A Review of the Research," *J. Real Estate Lit.*, vol. 2, no. 2, pp. 13–45, 1994, https://doi.org/10.1080/10835547.1994.12090037.
2. D. J. Pauline, and D. Selvi, "An Analytical View of Hypermarkets vs Supermarkets," *J. Manag. Res. Anal.*, vol. 05, no. 2(1), pp. 157–160, 2018, [Online]. Available: http://jmraonline.
3. A. Duvenhage, and A. Kruger, "Retail Format Selection: Occupancy Cost and Profitability," *J. Econ. Financ. Sci.*, vol. 10, no. 2, pp. 356–375, 2017, https://doi.org/10.4102/jef.v10i2.21.
4. H. Erbıyık, S. Özcan, and K. Karaboğa, "Retail Store Location Selection Problem with Multiple Analytical Hierarchy Process of Decision Making an Application in Turkey," *Procedia—Soc. Behav. Sci.*, vol. 58, pp. 1405–1414, 2012, https://doi.org/10.1016/j.sbspro.2012.09.1125.
5. Y. K. A. A. Migdadi, and E. H. A. Abdel-Rahman, "The Impact of Grocery Retail Superstores Location on the Customer's Perceived Quality of Servicescape in Jordan," *Qual. Manag. J.*, vol. 27, no. 2, pp. 120–132, 2020, https://doi.org/10.1080/10686967.2020.1722045.
6. J. L. Warren, and P. Gordon-Larsen, "Factors Associated with Supermarket and Convenience Store Closure Author (s): Joshua L. Warren and Penny Gordon-Larsen Published by: Wiley for the Royal Statistical Society Stable," *J. R. Stat. Soc. Ser. A (Statistics Soc.)*, vol. 181, no. 3, pp. 783–802, 2018, https://www.jstor.org/stable/10.2307/48547514 REFERENCES Linked r.
7. T. Tlapana, "The Impact of Store Layout on Consumer Buying Behaviour: a Case of Convenience Stores from a Selected Township in Kwazulu Natal," *Int. Rev. Manag. Mark.*, vol. 11, no. 5, pp. 1–6, 2021, https://doi.org/10.32479/irmm.11583.
8. Y. E. Kim, "A Study on Performance of Retail Store: Focused on Convenience Store," *J. Distrib. Sci.*, vol. 15, no. 2, pp. 47–51, 2017, https://doi.org/10.15722/jds.15.2.201702.47.
9. H. Zhang, Y. Hou, H. Feng, and C. Xu, "Solving the Location Selection Problem of Self-Service Stores from the Perspectives of Sustainability and Uncertainty," *J. Environ. Public Health*, vol. 2022, p. 1518755, 2022, https://doi.org/10.1155/2022/1518755.
10. S. Varshney, J. Raval, R. Priyadarshi, V. Manek, Y. Dadheech, and M. Shah, "Optimal Green Energy Estimation for Sustainable Electric Vehicle Networks Using Queueing-Based Retrial Process," *1st Int. Conf. Sustain. Energy Technol. Comput. Intell. Towar. Sustain. Energy Transition, SETCOM 2025*, pp. 0–5, 2025, https://doi.org/10.1109/SETCOM64758.2025.10932417.

11. S. Varshney *et al.*, "Stochastic Modeling of Electric Vehicle Infrastructure Using Queueing-Theoretical Approach," *Results Eng.*, vol. 25, no. October 2024, p. 104149, 2025, https://doi.org/10.1016/j.rineng.2025.104149.
12. K. Peng, and N. Kaza, "Availability of Neighbourhood Supermarkets and Convenience Stores, Broader Built Environment Context, and the Purchase of Fruits and Vegetables in US Households," *Public Health Nutr.*, vol. 22, no. 13, pp. 2436–2447, 2019, https://doi.org/10.1017/S1368980019000910.
13. R. Priyadarshi, and S. Routroy, "Vertical Integration Level Selection for Value Addition of Herbal Products: A Farmer's Perspective," *Mater. Today Proc.*, vol. 5, no. 9, pp. 18354–18361, 2018, https://doi.org/10.1016/j.matpr.2018.06.174.
14. R. Priyadarshi, A. Panigrahi, S. Routroy, and G. K. Garg, "Demand Forecasting at Retail Stage for Selected Vegetables: A Performance Analysis," *J. Model. Manag.*, vol. 14, no. 4, pp. 1042–1063, 2019, https://doi.org/10.1108/JM2-11-2018-0192.
15. R. Lalchandani, A. Chitnis, and R. Priyadarshi, *Mix Method Assessment of Sustainable Consumption Preferences towards Omni-Channel Operations*, vol. 2024, no. Icetsbp. Atlantis Press International BV, Dordrecht, 2024, https://doi.org/10.2991/978-94-6463-544-7.

Chapter 9

Formulation and Optimal Solution of a Fuzzy Decision-Making Linear Model with Application in Healthcare

Anuradha Sahoo and A. P. Maharana

9.1 INTRODUCTION

One of the uncertainties associated with managing healthcare resources, such as beds, staff, and equipment, is the uncertainty of patient demand. When data is uncertain, such as when scheduling staff shifts, linear fuzzy optimization aids in decision-making. Fuzzy linear optimization manages the relationships between resources and results in more complex situations, guaranteeing more effectiveness, lower expenses, and better patient care. Some major contributions by the authors in this context are given below.

Khan et al. [1] suggested a hybrid fuzzy decision tree model for cancer prediction. Results used in the SEER breast cancer dataset demonstrate that the fuzzy technique performs better in terms of balance and robustness than standard crisp classifiers. A fuzzy multi-criteria group decision-making framework was presented by Dursun et al. [2] for disposing of medical waste. Economic, technological, environmental, and social factors are taken into account when evaluating options such as disinfection and incineration using fuzzy set theory. Mezei et al. [3] proposed a fuzzy optimization model with chance-constrained optimization to study mobile health and wellness apps. Using an example of walking pace optimization, a novel method for dealing with interval-valued fuzzy integers and optimizing user behavior for improved health outcomes was presented.

Mardani et al. [4] examined traditional and fuzzy decision-making methods in the healthcare industry. With an emphasis on service quality assessment, it categorized the research into nine application areas and identified analytic hierarchy process (AHP) and hybrid approaches as the most widely utilized techniques. Ozsahin et al. [5] examined the application of fuzzy logic, fuzzy multicriteria decision analysis, and fuzzy linear programming (FLP) in medicine. Fuzzy logic helped with activities like patient monitoring, treatment selection, and disease diagnosis. Afshari et al. [6] used the Delphi method to present a fuzzy linguistic assessment strategy for choosing hospital management. In a case study studied in Mashhad, Iran, a fuzzy

 DOI: 10.1201/9781003642091-9

decision-making model was created and evaluated to show its potential and efficiency in practical settings.

Shie et al. [7] proposed a structural design decision model, resulting in the use of fuzzy approaches. It identifies reasons, ranks risks, examines service failure points, and comes up with creative fixes. The model's ability to improve service design and delivery was confirmed by empirical research conducted in Beijing. Lu et al. [8] examined large-scale group decision-making in the healthcare industry and used bibliometric analysis to identify structures and issues. Al Mohamed et al. [9] suggested a hybrid fuzzy multi-criteria decision model that combines fuzzy technique for order preference by similarity to ideal solution (FTOPSIS) and fuzzy analytic hierarchy process (FAHP). While FTOPSIS ranks alternatives, FAHP focused on criteria, demonstrating how well they handle location selection difficulties. Seyfi-Shishavan et al. [10] proposed a fuzzy multi-period mathematical model for optimizing blood supply chain networks in the event of natural disasters. To reduce expenses and shortages, the model takes into account factors such as supply, demand, and failure rates.

Pamucar et al. [11] proposed a unique decision-making technique combining measuring attractiveness and a new distance-based assessment method under fuzzy rough numbers used to handle the supplier selection problem for medical face masks and shields during COVID-19. Quasim et al. [12] suggested the single-stage healthcare model with fuzzy decision-making model, which used a fuzzy neural network for healthcare prediction and fuzzy AHP-TOPSIS to evaluate technological integration and data privacy. Compared to current models, the results suggest better risk assessment performance. Goodarzian et al. [13] suggested a fuzzy inference method. When establishing detoxification facilities, it seeks to reduce costs and the negative effects on the environment.

Albahri et al. [14] used medical and sociodemographic characteristics to prioritize patients with autism spectrum disorder across three urgency levels using a fuzzy multi-criteria decision-making technique. Gou et al. [15] used a unique approach using linguistic preference orderings to assess the allocation of medical health resources in public health emergencies. To prove that the effectiveness of the proposed method, it is used in a real-world scenario and contrasted with other approaches. Gul et al. [16] assigned nursing staff and bed capacity in a dental hospital. This study analyzed the use of transportation models with a fuzzy mixed integer linear programming approach. The results indicate that the final 3 months of the year had the highest patient demand, requiring more capacity.

This work contributed a new way to improve healthcare by combining fuzzy logic with linear programming. It manages uncertainties in resource allocation and patient care, for better planning and reducing expenses. The α-cut technique and graded mean integration (GMI) method convert fuzzy data into crisp data. By using LINGO software, the model provides optimal solutions for a real healthcare system. It helps decision-makers use their resources more efficiently while improving patient care in hospitals.

The structure of this chapter is as follows: an introduction and a summary of the key references are provided in Section 9.1. In Section 9.2, some preliminary definitions are stated. The complete formulation of the problem is considered in Section 9.3. In Section 9.4, the methodology is discussed. Section 9.5 presents a numerical example to demonstrate the proposed model. Finally, a brief conclusion is stated in Section 9.6.

9.2 BASIC PRELIMINARIES

Definition 9.2.1 (Decision-Making Problem): The process of choosing the best option from a range of alternatives is a component of decision-making problems. The essential elements of decision-making challenges are (i) alternatives (available possibilities criteria), (ii) elements (that are used to assess the possibilities), (iii) objectives (desired results), and (iv) constraints (limitations).

Definition 9.2.2 (Optimization Problem): Finding the optimal solution from a range of workable options that, given limitations, maximize or minimize a goal is known as an optimization problem. The main components are (i) objective function, (ii) decision variables, (iii) constraints, and (iv) feasible region.

Definition 9.2.3 (Fuzzy Set): A fuzzy set is a type of group or collection where the items can belong to it partially, rather than fully or not at all. Each item has a membership value between 0 and 1 that shows how much it belongs to the group. If X is a collection of objects, then fuzzy set $\tilde{A}$ in X is a set of ordered pairs given by

$$\tilde{A} = \left\{\left(x, \mu_{\tilde{A}}(x)\right) | x \in X\right\}.$$

Definition 9.2.4 (Membership Function): For a fuzzy set $\tilde{A}$ defined on a set X, the membership function $\mu_{\tilde{A}} : x \to [0,1]$ is a function that assigns to each element $x \in X$.

Definition 9.2.5 (Fuzzy Number): One kind of fuzzy set that reflects ambiguous numerical values is a fuzzy number. A fuzzy set $\tilde{A}$ on R must have the following characteristics to be a fuzzy number.

- Normality
- Convexity
- Piecewise continuity
- Boundedness

Definition 9.2.6 (α-Cut of Fuzzy Number): For a fuzzy number $\tilde{A}$ defined on R with a membership function $\mu_{\tilde{A}}(x)$, the α-cut is defined as

$\tilde{A}_\alpha = \{x \in R | \mu_A(x) \geq \alpha\}$, where $\alpha \in [0,1]$. The $x \in [L_\alpha, U_\alpha]$ α-cut can also be written as

$\tilde{A}_\alpha = [L_\alpha, U_\alpha]$, where L_α is the lower bound and U_α is the upper bound.

Definition 9.2.7 (Triangular Fuzzy Number):

A triangular fuzzy number is characterized by a triangular-shaped membership function. It is defined by three parameters: *a*, *b*, and *c*. The membership function of the triangular fuzzy number can be defined as

$$\mu_A(x) = \begin{cases} \dfrac{x-a}{b-a}, a \leq x \leq b \\ \dfrac{c-x}{c-b}, b \leq x \leq c \\ 0, otherwise \end{cases}$$

Definition 9.2.8 (α-Cut of Triangular Fuzzy Number):

For a given α, the α-cut of the triangular fuzzy number $\tilde{A}$, denoted as $\tilde{A}_\alpha$, is the set of values $x \in [L_\alpha, U_\alpha]$, where $\mu_{\tilde{A}}(x) \geq \alpha$. The lower bound $L_\alpha = a + \alpha(b-a)$ and the upper bound $U_\alpha = c - \alpha(c-b)$. Thus, the α-cut of the triangular fuzzy number is $\tilde{A}_\alpha = [a + \alpha(b-a), c - \alpha(c-b)]$.

Definition 9.2.9 (Fuzzy Arithmetic Operation): Let $\tilde{A}_\alpha = \left[L_\alpha^A, U_\alpha^A\right]$ and $\tilde{B}_\alpha = \left[L_\alpha^B, U_\alpha^B\right]$. Then the fuzzy arithmetic operations are as follows:

Addition: $\tilde{A}_\alpha + \tilde{B}_\alpha = \left[L_\alpha^A + L_\alpha^B, U_\alpha^A + U_\alpha^B\right]$

Subtraction: $\tilde{A}_\alpha - \tilde{B}_\alpha = \left[L_\alpha^A - U_\alpha^B, U_\alpha^A - L_\alpha^B\right]$

$$\text{Multiplication: } \tilde{A}_\alpha . \tilde{B}_\alpha = \begin{bmatrix} \min\left(L_\alpha^A L_\alpha^B, L_\alpha^A U_\alpha^B, U_\alpha^A L_\alpha^B, U_\alpha^A U_\alpha^B\right), \\ \max\left(L_\alpha^A L_\alpha^B, L_\alpha^A U_\alpha^B, U_\alpha^A L_\alpha^B, U_\alpha^A U_\alpha^B\right) \end{bmatrix}$$

$$\text{Division: } \tilde{A}_\alpha . \tilde{B}_\alpha = \begin{bmatrix} \min\left(L_\alpha^A / L_\alpha^B, L_\alpha^A / U_\alpha^B, U_\alpha^A / L_\alpha^B, U_\alpha^A / U_\alpha^B\right), \\ \max\left(L_\alpha^A / L_\alpha^B, L_\alpha^A / U_\alpha^B, U_\alpha^A / L_\alpha^B, U_\alpha^A / U_\alpha^B\right) \end{bmatrix},$$

where $0 \notin \left[L_\alpha^B, U_\alpha^B\right]$.

9.3 MODEL FORMULATION

In this section, we propose a fuzzy decision-making linear model with application in healthcare. When there is uncertainty in the data that frequently occurs in the healthcare industry, FLP is a technique used to make decisions.

The options to be made, such as how many beds to assign, how many staff to assign, or how to schedule treatments, are represented by the decision variables $(x_1, x_2, \ldots, x_n)$.

The purpose of the decision-making process is known as the objective function. For example, we might want to minimize costs, improve efficiency, or provide better patient care. In FLP, the objective function or the constraints have fuzzy parameters. Constraints are the limits or rules, such as the number of doctors available or maximum patient wait time.

Now we formulate a model corresponding to a real-life situation. Suppose a hospital in the United States wants to allocate its limited ICU beds to two types of patients: critical patients (P1) and noncritical patients (P2), minimizing overall resource costs while ensuring care for all. The hospital faces uncertainties in both resource requirements and available resources, which are modeled as fuzzy constraints. For model formulation, the following parameters and variables are used.

Let

x_1: Number of ICU beds allocated to P1.
x_2: Number of ICU beds allocated to P2.
C_1: Cost coefficients of ICU beds for P1.
C_2: Cost coefficients of ICU beds for P2.
$\tilde{B}$: Fuzzy total ICU beds available.
$\tilde{D}_1$: Fuzzy P1 demand.
$\tilde{D}_2$: Fuzzy P2 demand.

The objective of the decision-maker is to minimize the overall costs of allocating ICU beds between P1 and P2 while considering uncertainties in resource availability and ensuring adequate care for both patient groups. So, the decision-making fuzzy linear programming problem (DMFLP) model is given as follows:

DMFLP: Minimize $Z = C_1 x_1 + C_2 x_2$

$$\text{subject to } x_1 + x_2 \le \tilde{B}$$

$$x_1 \ge \tilde{D}_1 ,$$

$$x_2 \ge \tilde{D}_2 ,$$

$$x_1, x_2 \ge 0 .$$

In the following section, a complete technique is discussed to solve the DMFLP model, and the corresponding optimal solutions are obtained.

9.4 METHODOLOGY

Since the DMFLP model is a fuzzy optimization model, obtaining an optimal solution to the problem is difficult. To solve the model, the deterministic equivalent of the model is obtained by the defuzzification approach. Initially, the α-cut technique is used to convert the fuzzy parameters to crisp interval, and then the GMI method is applied to convert the crisp interval into a single crisp value.

Let $\tilde{A}_\alpha$ be a fuzzy number. Its α-cut at level α is $\tilde{A}_\alpha = [L_\alpha, U_\alpha]$, where L_α is the lower bound at $-$ and U_α is the upper bound at α.

9.4.1 Graded Mean Integration

This method calculates the defuzzified value by averaging the contributions of each α-cut across the entire range of α from 0 to 1. The α-cut at any level α gives an interval $[L_\alpha, U_\alpha]$. Then the GMI formula for the fuzzy number $\tilde{A}$ using its α-cut is

$$\text{GMI}(\tilde{A}) = \int_0^1 \frac{L_\alpha + U_\alpha}{2} d\alpha.$$

Let's consider a triangular fuzzy number $\tilde{A} = (a, b, c)$. The α-cut is given by $\tilde{A}_\alpha = [a + \alpha(b - a), c - \alpha(c - b)]$.

The GMI formula becomes $\text{GMI}(\tilde{A}) = \int_0^1 \frac{a + \alpha(b - a) + c - \alpha(c - b)}{2} d\alpha$.

Solving the integral, we get

$$\text{GMI}(\tilde{A}) = \frac{a + 2b + c}{4}.$$

By using α-cut and GMI, the fuzzy parameters $\tilde{B}$, $\tilde{D}_1$, and $\tilde{D}_2$ presented in the DMFLP model are converted to crisp values as B, D_1, and D_2, respectively. Thus, the deterministic equivalent of DMFLP is given by the following decision-making crisp linear programming problem (DMCLP) model.

$$\begin{aligned}
\textbf{DMCLP: } & \text{Minimize } Z = C_1 x_1 + C_2 x_2 \\
& \text{subject to } x_1 + x_2 \le B, \\
& x_1 \ge D_1, \\
& x_2 \ge D_2, \\
& x_1, x_2 \ge 0.
\end{aligned}$$

Now the above DMCLP model is a general LPP that can be solved by using any LPP technique. Here we obtained the optimal solution by using LINGO software.

9.5 NUMERICAL EXAMPLE

In this section, we introduce a numerical example to examine the applicability of the model under the consideration of the following parameters.

Suppose a hospital in the United States wants to allocate its limited ICU beds to two types of patients: critical patients (P1) and noncritical patients (P2), minimizing overall resource costs while ensuring care for all. The hospital faces uncertainties in both resource requirements and available resources, which are modeled as fuzzy constraints. For P1, ICU bed costs \$500 per bed, and for P2, ICU bed costs \$300 per bed. Fuzzy total ICU beds available is $\tilde{B} = (80,90,100)$, fuzzy critical patient demand is $\tilde{D}_1 = (50,55,60)$, and fuzzy noncritical patient demand is $\tilde{D}_2 = (20,25,30)$.

Let the DMFLP model corresponding to the above data be given by DMFLP (1).

$$\text{Minimize } Z = 500x_1 + 300x_2$$

$$\textbf{DMFLP (1)}: \text{subject to } x_1 + x_2 \le \tilde{B},$$

$$x_1 \ge \tilde{D}_1,$$

$$x_2 \ge \tilde{D}_2,$$

$$x_1, x_2 \ge 0.$$

The α-cuts of $\tilde{B}$, $\tilde{D}_1$, and $\tilde{D}_2$ are [87,93], [53.5,56.5], and [23.5,26.5], respectively.

Now by using the GMI method and for $\alpha = 0.7$ the crisp values are 90, 55, and 25, respectively.

Thus, the above DMFLP (1) is converted to DMCLP (1) by using the α-cut technique and the GMI method and is given as follows:

$$\textbf{DMCLP (1)}: \text{Minimize } Z = 500x_1 + 300x_2$$

$$\text{Subject to } x_1 + x_2 \le 90,$$

$$x_1 \ge 55,$$

$$x_2 \ge 25,$$

$$x_1, x_2 \ge 0.$$

The above DMCLP (1) is solved by any LPP technique. Here, we used LINGO software and obtained the following optimal solutions.

$x_1 = 55$ (critical patient demand) and $x_2 = 25$ (noncritical patient demand).

Total number of beds: $x_1 + x_2 = 80$ (less than the available 90 beds, so a feasible solution).

Objective value (cost): $Z = 35,000$

Now the decision-maker can decide by observing that the optimal solution is to allocate 55 ICU beds to critical patients and 25 ICU beds to noncritical patients, with a total cost of \$35,000.

In the following section, we give a brief conclusion.

9.6 CONCLUSIONS

The fuzzy optimization model successfully reduced the expenses of resource allocation costs while handling all uncertainties in ICU beds. It provides a flexible and effective method for managing healthcare resources while ensuring both critical and noncritical patients receive the care they need. The fuzzy approach allows flexibility in decision-making. This work can be

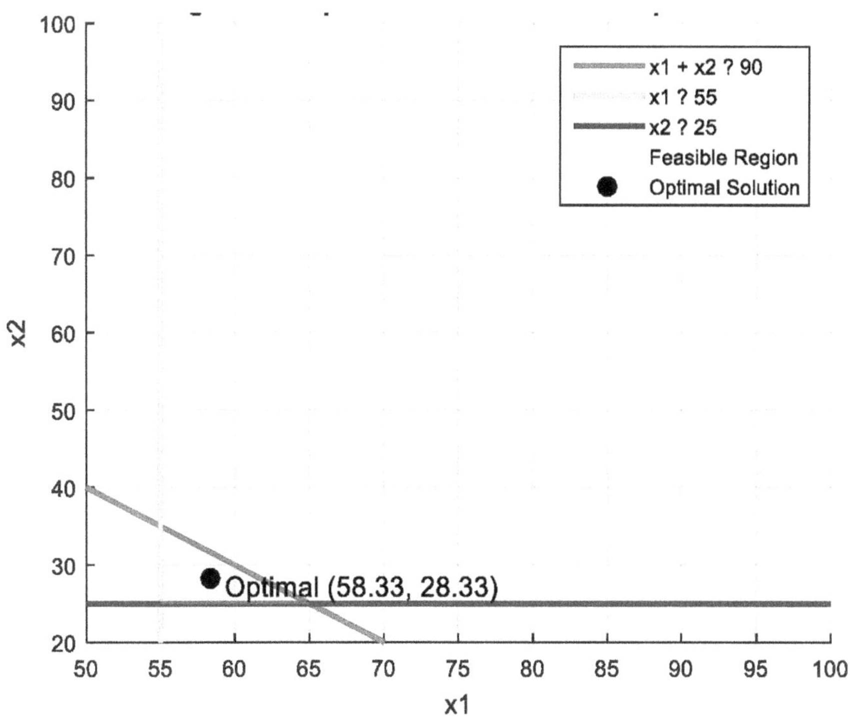

Figure 9.1 Feasible region and optimal solution of the linear optimization model.

extended to other mathematical programming problems to handle complex healthcare systems, namely, nonlinear programming problems and multi-objective programming problems.

REFERENCES

1. Khan, M. U., Choi, J. P., Shin, H., & Kim, M. (2008, August). Predicting breast cancer survivability using fuzzy decision trees for personalized healthcare. In *2008 30th Annual International Conference of the IEEE Engineering in Medicine and Biology Society* (pp. 5148–5151). IEEE.
2. Dursun, M., Karsak, E. E., & Karadayi, M. A. (2011). A fuzzy multi-criteria group decision-making framework for evaluating healthcare waste disposal alternatives. *Expert Systems with Applications*, *38*(9), 11453–11462.
3. Mezei, J., & Nikou, S. (2018). Fuzzy optimization to improve mobile health and wellness recommendation systems. *Knowledge-Based Systems*, *142*, 108–116.
4. Mardani, A., Hooker, R. E., Ozkul, S., Yifan, S., Nilashi, M., Sabzi, H. Z., & Fei, G. C. (2019). Application of decision making and fuzzy sets theory to evaluate the healthcare and medical problems: A review of three decades of research with recent developments. *Expert Systems with Applications*, *137*, 202–231.
5. Ozsahin, D. U., Uzun, B., Ozsahin, I., Mustapha, M. T., & Musa, M. S. (2020). Fuzzy logic in medicine. In *Biomedical Signal Processing and Artificial Intelligence in Healthcare* (pp. 153–182). Academic Press, Cambridge.
6. Afshari, A. R., Moghaddam, A. B., & Khorsand, M. (2021). Fuzzy Hospital Manager Selection. *Engineering Management and Competitiveness (EMC 2021)*, 3.
7. Shie, A. J., Lee, C. H., Yu, S. Y., & Wang, C. (2021). A fuzzy design decision model for new healthcare service conceptualization. *International Journal of Fuzzy Systems*, *23*, 58–80.
8. Lu, K., & Liao, H. (2022). A survey of group decision-making methods in Healthcare Industry 4.0: bibliometrics, applications, and directions. *Applied Intelligence*, *52*(12), 13689–13713.
9. Al Mohamed, A. A., Al Mohamed, S., & Zino, M. (2023). Application of fuzzy multicriteria decision-making model in selecting pandemic hospital site. *Future Business Journal*, *9*(1), 14.
10. Seyfi-Shishavan, S. A., Donyatalab, Y., Farrokhizadeh, E., & Satoglu, S. I. (2023). A fuzzy optimization model for designing an efficient blood supply chain network under uncertainty and disruption. *Annals of Operations Research*, *331*(1), 447–501.
11. Pamucar, D., Torkayesh, A. E., & Biswas, S. (2023). Supplier selection in healthcare supply chain management during the COVID-19 pandemic: a novel fuzzy rough decision-making approach. *Annals of Operations Research*, *328*(1), 977–1019.
12. Quasim, M. T., Shaikh, A., Shuaib, M., Sulaiman, A., Alam, S., & Asiri, Y. (2023). *Fuzzy Decision-Making Method Based Evaluation of Smart Healthcare*

Management, 03 January 2023, PREPRINT (Version 1) available at Research Square. https://doi.org/10.21203/rs.3.rs-1504815/v1.
13. Goodarzian, F., Ghasemi, P., Gunasekaran, A., & Labib, A. (2024). A fuzzy sustainable model for COVID-19 medical waste supply chain network. *Fuzzy Optimization and Decision Making*, *23*(1), 93–127.
14. Albahri, A. S., Hamid, R. A., Alzubaidi, L., Homod, R. Z., Zidan, K. A., Mubark, H., . . . & Alamoodi, A. H. (2024). Prioritizing complex health levels beyond autism triage using fuzzy multi-criteria decision-making. *Complex & Intelligent Systems*, 1–30.
15. Gou, X., Xu, X., Deng, F., Zhou, W., & Herrera-Viedma, E. (2024). Medical health resources allocation evaluation in public health emergencies by an improved ORESTE method with linguistic preference orderings. *Fuzzy Optimization and Decision Making*, *23*(1), 1–27.
16. Gul, A. Y., & Karadayi-Usta, S. (2024). Fuzzy transportation model for resource allocation in a dental hospital. In *Decision Making in Healthcare Systems* (pp. 151–171). Springer International Publishing, Cham.

Chapter 10

Multi-Objective Fuzzy Programming Problems with Triangular Fuzzy Parameters Using the Signed Distance Method

Rashmi Ranjan Ota

10.1 INTRODUCTION

Linear programming is one of the most prevalent operations research techniques for resolving real-world problems. A linear objective function and a set of constraints derived from linear equalities and inequalities characterize this type of mathematical programming. Applications of linear programming are frequently used in several industries to optimize the use of raw materials. Optimization problems involving engineering applications frequently necessitate the accomplishment of many goals. These programming problems are referred to as multi-objective optimization. There are many different approaches to solving multi-objective problems since they are inherently conflicting. The ε-constraint technique is one of them, and it was first introduced by Haimes et al. [1]. Following the procedures of Haimes, Nikas et al. [2] introduced a robust augmented ε-constraint technique to locate exact solutions to multi-objective linear programming problems. Using the ε-constraint technique, Mesquita-Cunha [3] presented three methods for the representation of multi-objective integer linear programming problems. Yu and Solvang [4] created a mathematical formulation to select a place for providing various facilities, including treatment plants, recycling facilities, and disposal sites, and to supply suitable technologies for hazardous waste treatment and route transportation, based on a multi-objective mixed integer programming approach. They used the enhanced ε-constraint technique to construct the Pareto curve of the multi-objective problem.

To create the best acceptable hazardous material for road-rail multimodal transportation plan, Sun [5] suggested a fuzzy multi-objective linear routing optimization model. The fuzzy model was first converted into a crisp model using Jimenez's fuzzy ranking approach and the fuzzy expected value method. Subsequently, the ε-constraint method is used to determine the Pareto front for the multi-objective problem. Allehashemi et al. [6] devised a fuzzy multi-objective optimization model for a closed-loop supply chain network and applied it to a mobile network in the Peel region of Ontario, Canada. Weighted sums and ε-constraint approaches are employed to find

 DOI: 10.1201/9781003642091-10

an effective solution after the model has been defuzzified using a fuzzy quality function deployment ranking method. Ammar and Khalifa [7] introduced a linear fractional programming problem with interval values that is multi-objective and has fuzzy parameters in both the objective functions and restrictions. The solution process for multi-objective fuzzy linear programming with a triangular membership function was demonstrated by Beena T. Balan [8] in 2016. However, in their study, Gupta et al. [9] showed how an interactive goal programming model might assist agricultural economists in optimally allocating various types of available resources for the agricultural production planning problem. They took into consideration a multi-objective, fuzzy agricultural production planning problem. Multi-objective intuitionistic fuzzy linear programming and its use in transportation models have been examined by Jana B. et al. [10].

Fuzzy multi-objective linear programming (FMLOP) problem in which both technological coefficient and resources are fuzzy with linear membership function was studied by Veeramani et al. [11]. Bharti and Singh [12] worked on intuitionistic fuzzy optimization to solve multi-objective linear programming problems. Luhandjula et al. [13] developed an approach for solving fuzzy multi-objective programming problems. In the year 1978, Zimmerman [14] showed how to find optimal solutions for fuzzy programming and linear programming problems with several objective functions.

The main goal of this study is to optimize a multi-objective fuzzy linear programming problem (MOFLPP) with coefficients as triangular fuzzy numbers. The problem was first changed into a crisp model by using the signed distance method and finally solved using the ε-constraint method. The remainder of this chapter is structured as follows: accompanying the introduction, the basic preliminaries regarding fuzzy sets and the mathematical formulation of the MOFLPP are given in Sections 10.2 and 10.3, respectively. The basic concept of the signed distance method is given in Section 10.4, and the ε-constraint method is given in Section 10.5. A numerical example and the corresponding result analysis are given in Sections 10.6 and 10.7, respectively. Finally, a brief conclusion related to the discussed topic is presented in Section 10.8.

10.2 PRELIMINARIES

10.2.1 Fuzzy Set

Fuzzy sets can be understood as an extension of classical set theory, which was introduced by Zadeh in the year 1965, published in a chapter entitled "Fuzzy Sets." It can be well understood with the help of set membership function. It essentially permits partial membership, which implies that it has elements with different levels of set membership. This helps us grasp how fuzzy sets and classical sets differ from one another. A fuzzy set consists of two components: the set itself, which is usually not empty, and the component that establishes membership. The value of a membership shows how much a member is part of the source set. Particularly, we are discussing a fuzzy set's membership function.

If S is a fuzzy set of the universe X, then the membership function is stated as $\mu_S(x)$ and is defined as $\mu_S(x): X \to [0,1]$, where

$$\mu_S(x) = 1, \text{ if } x \text{ is totally in } S,$$

$$\mu_S(x) = 0, \text{ if } x \text{ is not in } S,$$

$$\mu_S(x) \in (0,1), \text{ if } x \text{ is partly in } S.$$

a. When X is a finite set $\{x_1, x_2, \ldots\ldots\ldots\ldots x_n\}$, a fuzzy set S on X is expressed as $S = \sum_{i=1}^{n} \mu_S(x_i)/x_i$, where $\mu_S(x_i), i = 1, 2, \ldots. n$ signifies the grade of membership of x_i in S.
b. When S is not finite, we can write $S = \int_X \mu_S(x)/x$.

10.2.2 Fuzzy Number

Fuzzy number is a special type of fuzzy set defined on the set of real numbers representing a range of possible values with associated degrees of membership. If a fuzzy set satisfies at least four requirements, such as it should be convex, its support should be bounded, the core S should have at least one point, and all α-cuts are closed intervals, then it will be a fuzzy number. The most common types of fuzzy numbers are triangular, trapezoidal, and pentagonal numbers. Fuzzy numbers are basically used where uncertain data is required in decision-making.

10.2.3 Triangular Fuzzy Number

Fuzzy numbers are available in different shapes. The most widely used fuzzy numbers among them is triangular fuzzy numbers. A triangular fuzzy number is defined by three points, such as a minimum value a_1, a most likely value a_2, and a maximum value a_3. It is represented as $A = (a_1, a_2, a_3)$.

This number is interpreted using a membership function (cf. Figure 10.1) as

$$\begin{aligned}
\mu_S(x) &= 0, x < a_1, \\
&= \frac{x - a_1}{a_2 - a_1}, a_1 < x < a_2, \\
&= \frac{a_3 - x}{a_3 - a_2}, a_2 < x < a_3, \\
&= 0, x > a_3
\end{aligned}$$

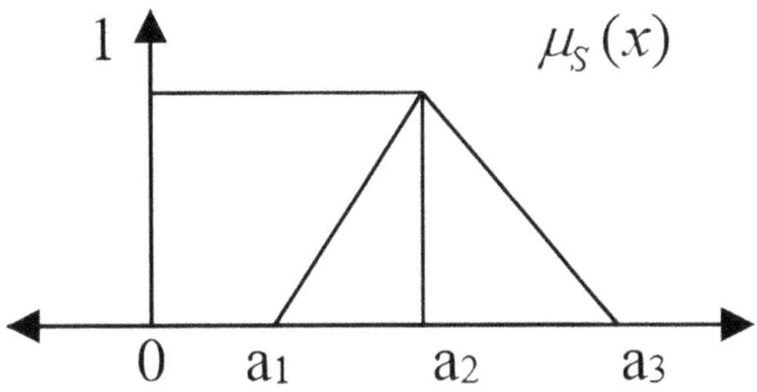

Figure 10.1 Triangular fuzzy number.

10.3 MATHEMATICAL FORMULATION OF FUZZY LINEAR PROGRAMMING PROBLEM

A general mathematical model of a MOFLPP with coefficients as triangular fuzzy numbers can be presented as follows:

$$\begin{aligned} &\min : \tilde{f}_p(x) = \sum_{t=1}^{T_p} \tilde{C}_{pt} \prod_{i=1}^{n} x_i, \quad p = 1,2,....,l \\ &\textit{subject to,} \\ &\tilde{d}_s(x) = \sum_{t=1}^{T_s} \tilde{C}_{st} \prod_{i=1}^{n} x_i \le 1, \quad s = 1,2,....,m. \\ &\qquad\qquad x_i > 0, \quad i = 1,2,.....,n \end{aligned} \tag{10.1}$$

where $\tilde{C}_{pt}$ and $\tilde{C}_{st}$ are triangular fuzzy parameters, i.e., $\tilde{C}_{pt} = \left(C^1_{pt}, C^2_{pt}, C^3_{pt}\right)$ and $\tilde{C}_{st} = \left(C^1_{st}, C^2_{st}, C^3_{st}\right)$.

T_p = Total number of terms in $\tilde{f}_p(x)$.

T_s = Total number of terms in the sth constraint.

10.4 SIGNED DISTANCE METHOD

It is used to defuzzify interval-valued fuzzy numbers into crisp numbers. If $\tilde{A} = (a,b,c)$ is a triangular fuzzy number, then the corresponding crisp value by the signed distance method is

$$\tilde{A} = \frac{a+2b+c}{4}.$$

If $\tilde{C}_{pt} = \left(C_{pt}^1, C_{pt}^2, C_{pt}^3\right)$ and $\tilde{C}_{st} = \left(C_{st}^1, C_{st}^2, C_{st}^3\right)$ are triangular fuzzy numbers, then they can be defuzzified by the signed distance method as follows:

$$\tilde{C}_{pt} = \left(C_{pt}^1, C_{pt}^2, C_{pt}^3\right) = \frac{C_{pt}^1 + 2C_{pt}^2 + C_{pt}^3}{4},$$

$$\tilde{C}_{st} = \left(C_{st}^1, C_{st}^2, C_{st}^3\right) = \frac{C_{st}^1 + 2C_{st}^2 + C_{st}^3}{4}.$$

Now model (10.1) can be written as

$$\begin{aligned}
&\min : \tilde{f}_p(x) = \sum_{t=1}^{T_p} \frac{C_{pt}^1 + 2C_{pt}^2 + C_{pt}^3}{4} \prod_{i=1}^{n} x_i, \quad p = 1,2,\ldots.,l \\
&subject\ to : \tilde{d}_s(x) = \sum_{t=1}^{T_s} \frac{C_{st}^1 + 2C_{st}^2 + C_{st}^3}{4} \prod_{i=1}^{n} x_i \le 1, \quad s = 1,2,\ldots.,m. \\
&x_i > 0, \quad i = 1,2,\ldots..,n
\end{aligned} \tag{10.2}$$

10.5 ε-CONSTRAINT METHOD

In order to find Pareto solutions for multi-objective problems, Haimes et al. [1] suggested the ε-constraint method in 1971. This strategy results in trade-off solutions between various objectives. Often, this approach is known as the trade-off approach. In this approach, only one objective function is optimized, keeping the others as inequality constraints. This concept is stated as follows:

$$\begin{aligned}
&\min : f_\gamma(x) \\
&subject\ to\ f_r(x) \le \varepsilon_r,\ r = 1,2,\ldots..,k\ \&\ r \ne \gamma. \\
&d_q(x) \le 0,\ q = 1,2,\ldots..,t \\
&x > 0 \in X
\end{aligned} \tag{10.3}$$

where X is the feasible region. The vector of upper bounds $\varepsilon = (\varepsilon_1, \varepsilon_2, \ldots\ldots\ldots.\varepsilon_k)$ defines the maximum value that each objective can have. One may vary the vector of upper bounds along the Pareto front in order to obtain a subset of the Pareto-optimal set for each objective and perform a new optimization process for each new vector. It may be displayed as

$L_r \le \varepsilon_r \le U_r,\ r = 1,2,\ldots..,k,$ where $L_r = \min f_r(x),\ \forall x \in X,\ r = 1,2,\ldots..,k$ and $U_r = \max f_r(x),\ \forall x \in X,\ r = 1,2,\ldots..,k$.

10.6 NUMERICAL EXAMPLE

$$\begin{aligned}
&\min : \overline{f_1}(x) = (1,2,3)x_1 + (2,4,6)x_2 + (1.5,2.5,3.5)x_3 \\
&\min : \overline{f_2}(x) = (0.5,1,1.5)x_2 \\
&\textit{subject to} \quad (0.7,1.2,2)x_1 + (1.5,2,2.3)x_2 = (2.2,2.8,3.2) \\
&\qquad\qquad (3.2,4,4.5)x_2 + (3.5,4,4.5)x_3 = (2.2,2.8,3.2) \\
&\qquad\qquad x_1, x_2, x_3 > 0
\end{aligned} \tag{10.4}$$

Let's convert triangular fuzzy numbers in Equation (10.4) into crisp values by using the signed distance method.

$$\begin{aligned}
&\min : f_1(x) = 2x_1 + 4x_2 + 2.5x_3 \\
&\min : f_2(x) = x_2 \\
&\textit{subject to} \quad 1.275x_1 + 1.95x_2 = 2.75 \\
&\qquad\qquad 3.925x_2 + 4x_3 = 2.75 \\
&\qquad\qquad \textit{where} \quad x_1, x_2, x_3 > 0
\end{aligned} \tag{10.5}$$

Optimal solution of the objective function $f_1(x)$

$$\begin{aligned}
&\min : f_1(x) = 2x_1 + 4x_2 + 2.5x_3 \\
&\textit{subject to} \quad 1.275x_1 + 1.95x_2 = 2.75 \\
&\qquad\qquad 3.925x_2 + 4x_3 = 2.75 \\
&\qquad\qquad x_1, x_2, x_3 > 0
\end{aligned}$$

The primal solution of the problem is obtained using LINGO software. For the decision values $x_1 = 1.085300, x_2 = 0.7006369, x_3 = 0,$ The optimal value of $f_1(x) = 4.973148$.

Optimal solution of the objective function $f_2(x)$

$$\begin{aligned}
&\min : f_2(x) = x_2 \\
&\textit{subject to} \quad 1.275x_1 + 1.95x_2 = 2.75 \\
&\qquad\qquad 3.925x_2 + 4x_3 = 2.75 \\
&\qquad\qquad x_1, x_2, x_3 > 0
\end{aligned}$$

The optimal solution to the problem obtained by LINGO is $f_2(x) = 0$ for $x_1 = 2.156863$, $x_2 = 0$, and

$$x_3 = 0.6875000.$$

Solution using the ε-constraint technique:

The main task is how to use the ε-constraint technique to resolve the original MOFLPP.

For $i = 1$ & 2, substitute the value of f_1 in f_2 and the value of f_2 in f_1 to find the lower bound L_i and the upper bound U_i of the functions f_i. The following is the outcome that was attained.

$$L_1 = 4.973148 \le (f_1 = \varepsilon_1) \le U_1 = 6.032479$$

$$L_2 = 0 \le (f_2 = \varepsilon_2) \le U_2 = 0.7006369$$

Solution of $f_1(x)$ by using the ε-constraint technique:

$$\begin{aligned} \min : f_1(x) &= 2x_1 + 4x_2 + 2.5x_3 \\ \textit{subject to} \quad & x_2 \le \varepsilon_2 \\ & 1.275x_1 + 1.95x_2 = 2.75 \\ & 3.925x_2 + 4x_3 = 2.75 \\ & x_1, x_2, x_3 > 0 \end{aligned}$$

Solution of $f_2(x)$ by using the ε-constraint technique:

$$\begin{aligned} \min : f_2(x) &= x_2 \\ \textit{subject to} \quad & 2x_1 + 4x_2 + 2.5x_3 \le \varepsilon_1 \\ & 1.275x_1 + 1.95x_2 = 2.75 \\ & 3.925x_2 + 4x_3 = 2.75 \\ & x_1, x_2, x_3 > 0 \end{aligned}$$

Table 10.1 Optimal Values of f_1.

ε_2	x_1	x_2	x_3	$f_1(x)$
0.1	2.003922	0.1	0.5893750	5.881281
0.3	1.698039	0.3	0.3931250	5.578891
0.5	1.392157	0.5	0.1968750	5.276501
0.6	1.239216	0.6	0.9875000E-01	5.125306
0.7	1.086275	0.7	0.6250000E-03	4.974112

Table 10.2 Optimal Values of f_2.

ε_1	x_1	x_2	x_3	$f_2(x)$
5.0	1.112462	0.6828774	0.1742655E-01	0.6828774
5.4	1.517082	0.4183181	0.2770253	0.4183181
5.8	1.921702	0.1537589	0.5366241	0.1537589
5.9	2.022857	0.8761905E-01	0.6015238	0.8761905E-01
6	2.124012	0.2147923E-01	0.6664235	0.2147923E-01

10.7 RESULT ANALYSIS

The signed distance method is a defuzzification technique in fuzzy logic that has advantages over other methods when working with non-isosceles triangular fuzzy numbers or in applications such as group opinion surveys. Its importance derives from its versatility in handling various fuzzy number shapes and its capacity to accelerate computations for ranking fuzzy numbers. In the course of research, a MOFLPP with all of the parameters represented by triangular fuzzy numbers is considered. Then, using the signed distance method, fuzzy parameters are converted to crisp values. Finally, the restructured problem is solved by the ε-constraint method. Since the problem taken is multi-objective, its solutions are conflicting in nature. There is no unique solution to this type of problem. Hence, a decision-maker has to make a decision on which value is appropriate to him/her.

10.8 CONCLUSION

In this study, MOFLPPs with triangular fuzzy parameters are examined. First, the signed distance approach is used to convert the MOFLPP into a multi-objective linear programming problem. The ε-constraint technique is then used to find a set of Pareto-optimal solutions. The computational results are obtained using LINGO software. This minor step in the direction of resolving MOFLPP could greatly aid researchers in their subsequent investigations.

REFERENCES

1. Haimes, Y. Y., Lasdon, L. S., & Wismer, D. A., On a Bicriterion formulation of problems integrated system identification and system optimization. *IEEE, Transactions on System, Man and Cybernatics*, *1*(3), 296–297, 1971.
2. Nikas, A., Fountoulakis, A., Forouli, A., & Doukas, H., A robust augmented ε-constraint method (AUGMECON-R) for finding exact solutions of multi-objective linear programming problems. *Operational Research*, 1–42, 2022.

3. Mesquita-Cunha, M., Figueira, J. R., & Barbosa-Póvoa, A. P., New ϵ-constraint methods for multi-objective integer linear programming: a Pareto front representation approach. *European Journal of Operational Research*, *306*(1), 286–307, 2023.
4. Yu, H., & Solvang, W. D., An improved multi-objective programming with augmented ε-constraint method for hazardous waste location-routing problems. *International Journal of Environmental Research and Public Health*, *13*(6), 548, 2016.
5. Sun, Y., A fuzzy multi-objective routing model for managing hazardous materials door-to-door transportation in the road-rail multimodal network with uncertain demand and improved service level. *IEEE Access*, *8*, 172808–172828, 2020.
6. Allehashemi, T., Amin, S. H., & Zolfaghari, S., A proposed multi-objective model for cellphone closed-loop supply chain optimization based on fuzzy QFD. *Expert Systems with Applications*, *210*, 118577, 2022.
7. Ammar, E. E., & Khalifa, H. A., Solving fully fuzzy multi-objective linear fractional programming problems based on fuzzy programming approach. *The Journal of Fuzzy Mathematics*, *27*(2), 301–311, 2019.
8. Beena, T., Solving multi-objective fuzzy linear optimization problem using fuzzy programming technique. *IOSR Journal of Mathematics*, *05*, 18–21, 2016.
9. Gupta, S., & Umaid, M., Multi-objective optimization in agricultural production planning with fuzzy parameters. *International Journal of Agricultural and Statistics Sciences*, *14*(1), 1–12, 2018.
10. Jana, B., & Roy, T. K., Multi objective intuitionistic fuzzy linear programming and its application in transportation model. *NIFS*, *13*, 1–18, 2007.
11. Veeramani, C., Duraisamy, C., & Nagoorgani, A., Solving fuzzy multi-objective linear programming problem with linear membership functions. Australian *Journal of Basic and Applied Sciences*, 2011.
12. Bharati, S. K., & Singh, S. R., Solving multi-objective linear programming problems using intuitionistic fuzzy environment optimization method: a comparative study. *International Journal of Modeling and Optimization*, *4*(1), 10–16, 2014.
13. Luhandjula, M. K., & Rangoaga, M. J., An approach for solving a fuzzy multi-objective programming problem. *European Journal of Operational Research*, *232*, 249–255, 2014.
14. Zimmerman, H. J., Fuzzy programming and linear programming with several objective functions. *Fuzzy Sets and Systems*, *1*, 45–55, 1978.

Chapter 11

Financial Risk and Corporate Valuation

Evidence from BSE 500 Listed Entities

Ashutosh Parhi, Kishor Chandra Meher, Rabi Narayan Subudhi, and Saroj Kumar Routray

11.1 INTRODUCTION

Financial risk management in the dynamic corporate finance environment has become more critical in maintaining the stability of firms, their profitability, and value in the long term. Risk management has become very important to contemporary firms due to the complexity of business operations, increased competition, and vulnerability of the business to both domestic and international economic uncertainties (Walker & Black, 2004; Varotto, 2011). Financial risk is uncertainty that surrounds the financial performances of firms and their valuation, and it can be broadly classified into two important dimensions, which include credit risk and market risk (Erasmus, 2008). Credit risk arises when parties fail to meet their agreed obligation and, thus, may lead to monetary losses, reduction in liquidity, and worsening of asset values (Holton, 2009). Excessive measure of credit risk has an adverse influence on the creditworthiness of a firm, on the expense of borrowing funds, and restricts access to capital markets, which can all damage firm value (Ghani & Nayan, 2016). On the other hand, market risk refers to the possibility of losses, which occur due to the adverse fluctuations in market variables such as interest rates, foreign exchange rates, stock prices, and commodity prices (Cruz, 2013). Unpredictability in such parameters may create disruption in the earnings, raising costs of capital, and reduce investor confidence (Daud et al., 2011; Erin et al., 2018).

The growing dynamism and uncertainty of the financial markets have compelled firms to be active and innovative in the manner in which they handle risks in an effort to promote their financial well-being and competitiveness (Imbierowicz & Rauch, 2014). The 2008–2009 world economic crisis and the financial crisis that occurred at the beginning of the twenty-first century made it clear that a strong system of enterprise risk management is necessary, which will be ready to evaluate and eliminate risk at all different levels of business activity (Rossi et al., 2013). In this regard, shareholder wealth is now a main agenda of corporate financial strategy due to effective risk management (Pagach & Warr, 2011). Firm value, which is a

DOI: 10.1201/9781003642091-11

significant measure of corporate performance, is mostly articulated in the market valuation of the securities of a firm. When the share price increases, the enterprise value and shareholder wealth increase, and vice versa. When the share price decreases, it means that the firm's value is lost, and the investor returns as well (Ganesh et al., 2013; Debby et al., 2014). Financial risk factors are important in the determination of firm value due to the impact it has on cash flows, the valuation of assets, and the judgment of investors (Ali et al., 2018). It has been empirically shown that well-developed risk management procedures, e.g., credit risk evaluation, hedging tactics, and market exposure anticipation, create stabilization of profits, risk safeguarding of profitability, and increased investor confidence (Beasley et al., 2008; Gordon et al., 2009; Matayo & Muturi, 2018). As examples, credit risk management can minimize defaults and bad debt that result in the long-term flow of income (Siringi & Obange, 2021), and market risk management helps avoid volatility, which provides more predictable financial results (Sisay, 2017).

Despite the extensive focus of past studies on the relationship between financial risk and profitability, a critical shortfall in scholarly work is evident concerning the impact of financial risk on firm value, especially among nonfinancial firms. Moreover, much of the existing evidence is concentrated within the banking and financial sectors, which operate under distinct regulatory and risk exposure frameworks. To fill this gap, the present study examines the credit risk and market risk effects on the enterprise value of nonfinancial companies in India based on the panel of enterprises that were listed in BSE 500 index between the years 2017 and 2023. The study helps in developing the existing body of literature by focusing on company valuation instead of profitability and enlarging empirical research to nonfinancial industries, thus offering a more detailed understanding of the importance of financial risk management in the process of generating corporate value.

The chapter has the following structure: Section 11.2 provides a thorough review of the literature and develops the hypotheses. Section 11.3 describes the research design and methodology adopted in the study. Section 11.4 discusses the empirical findings. Section 11.5 provides a commentary on the major findings. Section 11.6 presents the robustness tests that were performed to prove the results. Finally, Section 11.7 provides a summary of the key findings and recommendations on the further investigation of the issue on a scholarly level.

11.2 THEORETICAL BACKGROUND, LITERATURE REVIEW, AND HYPOTHESIS DEVELOPMENT

11.2.1 The Theoretical Background

This study draws on two key theories: modern portfolio theory (MPT) and the capital asset pricing model (CAPM), both of which offer essential frameworks for understanding how credit risk and market risk individually influence firm value.

The MPT put forward by Markowitz (1952) takes an organized way of approaching portfolios, and its concept of focus is diversification to maximize risk-adjusted yields. Within the corporate finance setting, the MPT implies that both market risk and credit risk influence the value of a firm by changing the general amount of risk in it. In particular, an increased level of credit risk, which means higher default probability, and market risk, which means a change in the interest rates and the value of equity stocks, will increase the volatility of the earnings of firms and their asset value. This volatility may affect some investors' perception and eventually firm valuation (Lintner, 1965). The MPT also places much emphasis on the correlation of assets in a portfolio, meaning that firms ought to use the interactions of different risks. An example is that being exposed to market risk can be solved by investing in assets that are less affected by market changes, and hence the risk is diversified in the portfolio. Also, among its derivatives, the MPT espouses the use of financial instruments to control such risks, earnings smoothness, and firm value preservation by using financial instruments such as interest swaps and credit default swaps. A study conducted by Bodie, Kane, and Marcus (2014) further puts weight on the importance of having a diversified risk management plan on firm value. In addition to that, Hubbard (1998) supports the notion that risk management can turn any possible risk into an opportunity for value adding. Hovakimian and Li (2011) confirm this opinion, and their research results state that companies that have well-established credit risk management systems can be characterized by improved corporate valuation levels.

On the other hand, the CAPM formulated by Sharpe (1964) brings direct association between market risk, which is the coefficient of the beta, and expected returns. The CAPM presumes that the returns of an asset are relative to the beta, which shows the sensitivity to the dynamics of the market. However, despite the fact that the CAPM gives an understanding of the risk-return trade-off formula, Fama and French (1992) studied the adequacy of beta with a view to establishing the variations in firm value. Their research had the implication that although there is a positive relationship between beta and returns, it does not explain all the factors that can lead to the valuation of firms. They mentioned other supplementary factors, such as company size and market ratio, as more significant factors of firm value. The result helps reflect the dimensions of the inadequacy of the CAPM in embracing the whole context of firm valuation determinants. To reinforce this argument, Banz (1981) showed that smaller firms often earned higher returns than larger firms even when beta was similar, suggesting that firm size and other characteristics may explain valuation better than beta alone. Furthermore, the study performed by Fama and French (2015) elaborated on the prior study, enriching it with the aspects of such variables as profitability and investment patterns and proving the position that the multifactor model may be better used to understand the valuation of the companies. It is a broader picture that substantiates the understanding that it is not only the beta that can be used to interpolate the shifts in the values of firms.

11.2.2 Literature Review and Hypothesis Development

Financial risk, especially credit risk and market risk, has elicited a great deal of scholarly interest in terms of its effect on firm valuation. Whereas many studies indicate that proper risk management can increase firm performance, the effect on enterprise value, which is measured by the value Tobin's Q, is a new direction of empirical investigation. The proportion of the market value of the assets of an enterprise to the value of the assets based upon current costs or value of the same asset as a new asset, i.e., the ratio of market value of enterprise assets to current costs, referred to as Tobin's Q, is a forward-looking indicator of the value of a firm. An optimal Tobin's Q is said to signal that shareholders expect a more impressive growth or a better use of the assets of the firm, whereas a lower Q is significantly related to inefficiency or pessimism in the market. In this regard, credit and market risks may increase or decrease a firm's valuation, respectively, according to the manner in which they are handled.

There is substantial literature to tend to believe that well-managed credit risk has an upward effect on firm value. As an example, Demirgüç-Kunt and Huizinga (2000) established that the valuation of firms was higher in the banking industry, where companies possessed strong credit risk management. Research conducted by Elton and Green (2003) on the corporate bond market indicated that, through strategic credit risk management, the confidence of investors and the valuation of a firm might be increased. Zhang and Zhang (2011) and Liu and Wang (2015) reported that it was the firms in emerging economies that proactively managed exposure to credit, and those firms resulted in better performance in the market and a higher level of Tobin's Q. In the same manner, Hovakimian and Li (2011) emphasized that proper rebalancing of capital structure in the face of high credit risk may result in valuation advantages. Kothari, Shu, and Wysocki (2009) also argued on the signaling impact of sound credit risk management on the value of the market. Post-crisis literature, such as Chien and Chen (2021), Ghosh and Nanda (2022), and Li and Zhang (2023), confirmed that credit risk mitigation during times of economic volatility preserved and, in some cases, increased firm value. As demonstrated by Nguyen and Tran (2023), further quantification of credit analytics in enhancing enterprise value is enhanced by the innovations in FinTech. This positive relationship, however, depends on good management. Uncontrollable or excessive credit risk has the value-destructive potential that is warned about in several studies. Initial research by Altman (1968) using the Z-score model proved that credit risk is abnormally linked to financial distress and bankruptcy, which negatively affects firm value. The decline in equity value given the increase in default risk was modeled in a theoretical sense by Merton (1974). Empirical evidence from Ohlson (1980) and Almeida and Philippon (2007) suggests that firms facing high credit risk and weak governance are more likely to experience financial distress and lower market valuation. Publications by Acharya et al. (2013) and Giesecke et al. (2011) demonstrated that a combination of excessive leverage and lax credit risk controls contributes to stronger adverse impacts on Tobin's Q.

The relationship between market risk, as typically captured by beta in the CAPM framework, and firm value has yielded more ambiguous results. As Sharpe (1964) explains, the expected returns are expected to be higher with a greater beta and may increase market valuation. Empirical studies have, however, contradicted this assumption. Litzenberger and Ramaswamy (1979) and Amihud and Mendelson (1986) discovered that market risk influences pricing because of the overwhelming effects of other factors, including liquidity, taxes, and dividend preferences. Campbell and Vuolevenaho (2004) took this argument a step further by breaking down beta into cash flow beta and discount rate beta, and revealed that the valuation relevance of market risk is more subtle than initially hypothesized. In other instances, the effect of market risk on Tobin's Q is statistically insignificant, particularly when industry forces and company-specific risks are prevalent. Ferson and Harvey (1991) affirmed that the effects of market risk are diverse in geographical locations and in different cycles of an economy. An example is the case of emerging markets, where volatility and institutional effects at the local level tend to surpass the effect of systematic risk in the market. More recent studies, including those by Jegadeesh and Titman (2019) and Ang et al. (2006), have pointed to behavioral and firm-specific variables, including but not limited to momentum and idiosyncratic volatility, that undermine the predictive capability of beta on firm value. According to sectoral studies by Heston et al. (1995), technological change, regulation, and competition may have an intermediate or even neutralizing effect on the impact of beta on the Tobin's Q.

- Hypothesis 1 (H1): Credit risk positively impacts firm value.
- Hypothesis 2 (H2): Market risk has a positive consequence on firm value.

11.3 RESEARCH FRAMEWORK

11.3.1 Sample Selection

This research study focuses on the contribution of financial risk, which is credit risk and market risk to firm value, using a panel data sample comprising Indian nonfinancial listed companies in the BSE 500 index in the period between 2017 and 2023. Financial data were collected through the

CMIE Prowess IQ database, which is a common and reputable source of firm-level data in India. A sample size of 500 firms was used, but due to systematic sampling, data consistency and reliability were achieved. Firms that had not yet submitted their financial information in 2 consecutive years were eliminated to ensure that there were balanced and continuous panel observations. Moreover, banking, financial services, and insurance (BFSI) companies were eliminated because of their unique capital structure, regulatory structure, and risk structure, which are very different from those of nonfinancial companies. Firms that have been in mergers or acquisitions as well as delisting within the sample were also filtered and sifted to eliminate distortive effects on the firm value estimates. Using these criteria, 284 nonfinancial firms were obtained as the final sample with 1,988 firm-year observations in total. Although these exclusions increase the strength of the data, the research recognizes that there is a possibility of the existence of survivorship bias because there might have been firms with an incomplete or volatile history of reporting.

11.3.2 Variables

11.3.2.1 Response Variable

Tobin's Q is a representation of firm value, which is the response variable of the study. Tobin's Q is measured as the market value of equity plus the book value of debt divided by total assets. A higher Tobin's Q reflects stronger market expectations regarding firm growth and asset utilization. The larger the Tobin's Q is, the more confidence the market has in the possibility of firm growth and effective use of assets, and the smaller the Q is, the more negative the market's attitude toward the company is, or the inefficient use of assets. The measure has been extensively known in the literature as a futuristic measure of firm valuation.

11.3.2.2 Explanatory Variable

The independent variables representing financial risk are divided into two dimensions: credit risk and market risk. Credit risk is quantified using the Altman Z-score, a composite indicator proposed by Edward Altman (1968) to assess the possibility of financial distress. The Altman Z-score is calculated using the formula: $Z = 1.2X_1 + 1.4X_2 + 3.3X_3 + 0.6X_4 + 1.0X_5$. In this model, X_1 represents the ratio of working capital to total assets, reflecting short-term liquidity; X_2 denotes retained earnings divided by total assets, indicating the firm's cumulative profitability; X_3 measures earnings before interest and taxes (EBIT) relative to total assets, serving as an indicator of operating efficiency; X_4 captures the market value of equity in relation to the book value of total liabilities, representing leverage and investor confidence; and X_5 corresponds to the sales-to-total-assets ratio, which assesses asset turnover and operational efficiency. A higher Z-score signifies better financial health and lower credit risk. Market risk is captured by the beta coefficient (β), which measures the sensitivity of a company's stock returns to changes in the overall market index. A higher beta indicates greater volatility and higher exposure to systematic market risk, while a lower beta suggests reduced sensitivity to market fluctuations.

11.3.2.3 Control Variables

To control for firm-level factors that may influence Tobin's Q independently of financial risk, the study includes three control variables. First, the company's size is determined by taking the natural logarithm of its total assets, following the methodology of Titman and Wessels (1988), and is intended to capture the effects of scale, diversification, and operational efficiency. Second, leverage is included as the ratio of long-term debt to total assets, reflecting the firm's reliance on debt financing, in line with Rajan and Zingales (1995). Leverage can have dual implications: while it may signal growth potential through capital access, excessive leverage can indicate heightened financial vulnerability. Third, enterprise capitalization is used as a control for market worth and investor perception. This variable, aligned with the approach of Demsetz and Villalonga (2001), includes market capitalization adjusted for cash holdings and debt obligations to represent a firm's net financial strength. The detailed descriptions are presented in Table 11.1.

11.3.2.4 Empirical Model

The study employs panel data regression as its methodological framework to examine the association between financial risk and business value. It proposes two regression models to analyze how market risk (β value) and credit risk (Altman Z-score) influence Tobin's Q.

$$\begin{aligned} Tobin's\ Q = {} & \beta_1\, Credit\ Risk + \beta_2\, Firm\ Size + \\ & \beta_3\, Leverage + \beta_4\, Enterprise\ Value \end{aligned} \tag{11.1}$$

$$\begin{aligned} Tobin's\ Q = {} & \beta_1\, Market\ Risk + \beta_2\, Firm\ Size + \\ & \beta_3\, Leverage + \beta_4\, Enterprise\ Value \end{aligned} \tag{11.2}$$

Credit risk is calculated using the Altman Z-score, while market risk is represented by the beta (β) value. Company size is calculated as the total assets' natural logarithm, and leverage is determined by the proportion of long-term debt to total assets. Enterprise capitalization is computed by adding market capitalization and total debt, and then subtracting cash and cash equivalents.

Table 11.1 Summary of Variables.

Indicators	*Explanation*	*Proxy*	*Symbol*	*Measurement*
Dependent Variable				
Firm value	An indicator of a firm's comprehensive performance and financial health, reflecting investor perceptions of its future growth potential	Tobin's Q	TQ	Market value of equity + book value of debt/total assets
Independent Variable				
Market Risk	A statistical indicator that shows the sensitivity of an enterprise's stock returns to changes in broader market movements.	Beta	β	Ratio of the covariance between a stock's returns and the market's returns.
Credit risk	An enterprise's ability to meet its debt obligations	Altman *Z*-score	*Z*	$Z = 1.2X_1 + 1.4X_2 + 3.3X_3 + 0.6X_4 + 1.0X_5$
Control Variables				
Firm size	Proxy for the company's operational scale and capacity	Total assets	FS	Log value of total assets
Leverage	Firm's reliance on debt relative to its total assets	Noncurrent liabilities and total assets	LV	Ratio of long-term liabilities to total assets
Enterprise capitalization	Indicator of the company's economic value and market position, incorporating debt obligations while excluding cash reserves	Market capitalization and debt	EC	Sum of market capitalization and total debt, minus cash and cash equivalents

Source: Author's compilation

11.4 EMPIRICAL ANALYSIS AND INTERPRETATIONS

11.4.1 Summary Statistics

Table 11.2 summarizes the descriptive statistics of all variables. The mean of Tobin's Q is 3.188, implying that there is a high dispersion of firm values. The mean credit risk is 4.111, which means that there are varying degrees of financial distress in firms. The mean of the market risk is 1.058, which indicates divergent responsiveness to market alterations. The mean size of firms is 4.822, which illustrates that there are huge variations in the sizes of the firms. The leverage is 0.141, and it varies in terms of debt financing. Finally, enterprise capitalization has an average of 397,586.9, indicating notable variation in the economic value of the firms under study.

11.4.2 Correlation Matrix and Variance Inflation Factor

Correlation coefficients and the variance inflation factor (VIF) of the variables used in this research can be found in Table 11.3. There is a positive and very high relationship (0.939) between Tobin's Q value, which is the measure of firm value and credit risk, indicating that as the level of credit risk increases, so does the value of Tobin's Q, whereby there is a possibility that it is the financial distress that affects the determination of the market price. Conversely, market risk indicates a negative correlation with Tobin's Q (−0.356), meaning that the stocks of the companies that are exposed to greater market fluctuations have lower valuation. There is also a negative correlation between firm size and Tobin's Q (−0.311), which means that larger firms could have an average of slightly lower market capitalization. Again, leverage is shown to have a negative correlation with Tobin's Q (−0.233), which implies that as the level of debt increases when measured against total assets, it may tend to influence a firm's value on a negative note. Conversely, the capitalization of enterprise is positively correlated with that

Table 11.2 Summary Statistics Result.

Attribute	*Obv.*	*Mean*	*St. Dv.*	*Min.*	*Max.*
Tobin's Q	1,988	3.18841	3.274405	.0367949	27.58642
Credit risk	1,988	4.110706	2.614673	−1.699814	21.56304
Market risk	1,988	1.058008	.3856041	.19	2.61
Firm size	1,988	4.822401	.608348	3.188394	6.987719
Leverage	1,988	.1414447	.129831	.000254	.6418816
Enterprise capitalization	1,988	397,586.9	1097,689	−418,563.2	1.88e+07

Source: Author's estimation

Note: Strongly balanced panel data from 2017 to 2023.

Table 11.3 Correlation Coefficients and VIF.

	Tobin's Q	*Credit Risk*	*Market Risk*	*Enterprise Size*	*Leverage*	*Enterprise Capitalization*	*VIF*
Tobin's Q	1						
Credit risk	0.938639	1					1.57
Market risk	−0.35562	−0.39301	1				1.20
Enterprise size	−0.31093	−0.35514	0.084847	1			1.82
Leverage	−0.23306	−0.3543	0.113922	0.350312	1		1.22
Enterprise capitalization	0.122531	0.098629	−0.13273	0.508672	0.098919	1	1.54
Mean value							1.47

Source: Author's estimation

of Tobin's Q (0.123), meaning that companies that are more capitalized will have higher valuations. VIF values are scrutinized in order to determine the existence of multicollinearity among independent variables. The findings have shown that the degree of multicollinearity is mild with a VIF of 1.47 and should be of no worry in this study.

11.5 REGRESSION OUTCOME

Table 11.4 presents the regression results. Market risk (Coef. = 1.224742) shows a statistically significant positive coefficient, indicating that firms exposed to higher market risk tend to have higher Tobin's Q values. This implies that investors are likely to see increased market risk as a chance of greater potential returns, and this raises the value of the firm. The statistics of credit risk (Coef. = 0.1777919) also have a statistically significant positive coefficient. This means that the larger the credit risk of a business, the larger the value of Tobin's Q. This means that effective management of credit risk would increase investor confidence and firm value. On the other hand, firm size (Coef. = –0.1062428) has a statistically significant negative coefficient. This hypothesizes that large businesses would tend to have a lower score of Tobin's Q. Larger businesses might experience drawbacks of decreased agility or increased administrative expenses, which might affect their perceived value relative to smaller competitors. The coefficient of leverage (= 2.906683) is statistically significant and positive, and the coefficient of enterprise capitalization (= 8.20e-08) is also statistically significant and positive. Higher leverage, which means that an enterprise is more dependent on debt financing, and larger enterprise values, which represent general economic value with both debts and market capitalization, are allied with higher values of the possible Tobin's Q. This shows the favorable effect of

Table 11.4 Results of the Regression Analysis.

	Explained Variable: Firm Value	
Explanatory Variable	*Estimated Coefficients*	*Std. Error*
Credit risk	1.224742*** (0.000)	.0115033
Market risk	.1777919*** (0.009)	.0681664
Firm size	–.1062428** (0.047)	.0533449
Leverage	2.906683*** (0.000)	.2049034
Enterprise capitalization	8.20e-08*** (0.003)	2.72e-08
_cons	–1.965648*** (0.000)	.2841423
Observations	1,988	
R^2	0.8932	

Source: As estimated by the author

p-values are shown in parentheses, with $^{**}p < 0.05$, $^{***}p < 0.001$, and $^{*}p < 0.10$.

financial structure and the firm's worth in accumulation by the investors in the firm value perception. The value of R^2 is 0.8932, which shows that 89.32% of the variance in Tobin's Q is explained by the explanatory variables in the model.

11.5.1 Panel Regression Results

The panel data regression analysis was conducted for a dataset comprising 284 Indian enterprises, as presented in Table 11.5. The study applied a combination of pooled ordinary least squares (OLS), fixed effects (FE), and random effects (RE) models to investigate the impact of financial risk on enterprise valuation. Prior to model estimation, the Breusch–Pagan Lagrange multiplier test was used to assess heteroskedasticity, and the test confirmed its presence $\left(\chi^2 = 199.39,\ p\text{-value} = 0.000\right)$, indicating the need for robust panel data techniques. The initial OLS regression (Table 11.4) was discovered to be biased by heterogeneity since the OLS model would not consider unobserved firm-specific characteristics that do not vary over time; therefore, it would produce potentially biased, inconsistent estimates of coefficients. Therefore, panel data models were used to overcome this shortcoming because they allow unobserved heterogeneity type of control, namely, time-invariant firm-specific variables that have the potential to affect firm value, e.g., managerial style, organization culture, or norms affecting the sector. The FE model was chosen because it is the most effective in controlling this firm-level heterogeneity since it is easy to control because all it does is to demean the data, i.e., it subtracts the within-firm average from each variable, thus eliminating the effect of all time-invariant factors. It then permits precise estimation of only the within-firm variation changes

Table 11.5 Panel Regression Outcomes—FE Model.

	Dependent Variable: Firm Value	
Explanatory and Control Variables	*Coefficient Values*	*Standard Error*
Credit risk	1.278545*** (0.000)	.0114227
Market risk	.1012746 (0.176)	.0749287
Firm size	.2316587*** (0.004)	.0794615
Leverage	2.546506*** (0.000)	.2213013
Enterprise capitalization	8.90e-08*** (0.003)	2.94e-08
_cons	−3.687183 (0.000)	.3968617
Observations	1,988	
R^2	0.8932	

Source: As estimated by the author

p-values are shown in parentheses, with $^{**}p < 0.05$, $^{***}p < 0.001$, and $^{*}p < 0.10$.

within time and makes the model especially useful where the panel data context might cause observable correlations of the explanatory variables and unobservable fixed characteristics. The RE model, conversely, is based upon the postulate that these firm-specific effects are independent of the regressors, which in empirical corporate finance may not be the case.

The Hausman test (Hausman, 1978) was used to formally identify the best model. The alternative argument favors the application of the FE model, whereas the null hypothesis holds that the RE model is reliable and effective. The test shows a χ^2 value of 35.91 (p-value = 0.00), which rejects the null hypothesis decisively and confirms that the FE model is a superior fit to the data as a consequence of the possibility of endogeneity between firm-specific effect and the explanatory variables. Model 1: The regression findings on the impact of credit risk on enterprise value were tested, whereas Model 2 determined the impacts of market risk. The good fit of the model is verified by the fact that R^2 is equal to 0.8932, meaning that the independent variables explain approximately 89.32% of the variance of Tobin's Q.

The output of the FE estimation is shown in Table 11.5. The diagnostic credit risk tool, Altman Z-score, has a statistically significant positive correlation with Tobin's Q (coefficient = 1.278545, $p < 0.001$). This implies that firms rated with a higher credit risk score, indicating greater financial well-being, are related to a higher market value. The result confirms Hypothesis 1 (H1), which means that effective credit risk management increases the value of the firm. Conversely, market risk as indicated by the beta coefficient of a firm has a positive, although statistically insignificant, association with Tobin's Q (coefficient = 0.1012746, $p = 0.176$). This indicates that changes in market risk do not always impact firm value in the sample, which results in the adoption of Hypothesis 2 (H2) with a qualification that market risk has a positive but insignificant association with company valuation. Control variables that are statistically significant include firm size, which has a positive value of 0.2316587 ($p = 0.004$), which means that the larger the firm, the higher the market values it. There is also a strong positive correlation between leverage and growth potential (coefficient = 2.546506, $p < 0.001$), indicating that an increase in debt can be an indication of more growth prospects or growth potential when administered correctly. Enterprise capitalization has a positive impact on firm value, too (coefficient = 8.90e-08, $p = 0.003$), which also supports the role of the capital strength in valuing the market.

11.6 ROBUSTNESS CHECK

In order to verify the strength and the generality of the main results, researchers adopt another indicator of firm value, i.e., price-to-book ratio (PBR) as the dependent variable, with the same specification of the FE regression

Table 11.6 Robustness Test Results.

	Dependent Variable: PBR	
Independent Variable	*Coefficient Values*	*Standard Error*
Credit risk	2.30709*** (0.000)	.0532965
Market risk	.2891693 (0.424)	.2891693
Firm size	1.168425*** (0.021)	1.168425
Leverage	16.65615*** (0.000)	16.65615
Enterprise capitalization	3.27e-07*** (0.019)	3.27e-07
_cons	−12.01089 (0.000)	−12.01089
Observations	1,988	
R^2	0. 6797	

Source: As estimated by the author

p-values are shown in parentheses, with $^{**}p < 0.05$, $^{***}p < 0.001$, and $^{*}p < 0.10$.

model. This is an alternative measure that complements the evaluation of Tobin's Q to show the relationship between firm value and book equity, giving this an alternative validation measure used in the empirical study. Table 11.6 presents the robustness test results, which reaffirm the primary analysis: credit risk (proxied by the Altman *Z*-score) maintains a statistically validated positive effect on enterprise value, reflected by a coefficient of 2.30709. This reinforces the conclusion that firms with stronger financial health or better-managed credit risk tend to exhibit higher market valuations. Conversely, market risk, proxied by beta, continues to exhibit a positive but statistically insignificant relationship with the PBR (coefficient = 0.2891693), mirroring its limited impact on the Tobin's Q model. The robustness model also reaffirms the significant influence of the control variables. Firm size (coefficient = 1.168425), leverage (coefficient = 16.65615), and enterprise capitalization (coefficient = 3.27e-07) all maintain statistically significant positive relationships with PBR, supporting the consistency of the earlier findings. An R^2 of 0.6797 indicates that approximately 67.97% of the variation in PBR is described by the independent variables, further validating the explanatory power of the model.

11.7 DISCUSSION AND FUTURE OUTCOMES

This study examines the impact of financial risk, which is credit risk and market risk, on the valuation of nonfinancial companies listed on the BSE 500 index between the period 2017 and 2023. It analyses 284 unique companies, with balanced 1,988 company-year observations. The companies chosen in the analysis are not financial institutions because they have sector-specific regulatory and risk characteristics. The main explanatory variables are credit risk, which is proxied by the Altman Z-score, and market risk, measured with the help of the beta coefficient. The dependent variable is Tobin's Q, a forward-looking measure of enterprise value that is well-established, and the control variables will include enterprise size, leverage, and enterprise capitalization. The diagnostics of multicollinearity were performed on the VIF, which has a mean of 1.47, which shows that there is no significant multicollinearity and that the estimates of coefficients can be considered robust. To consider the firm-level unobserved heterogeneity, the FE regression model was used, and the Hausman test showed that it was more appropriate than the RE model. The FE model gave a high R^2 value of 0.8932, which indicates high variation; i.e., 89.32% of the firm value can be explained by the explanatory variables in the model. Regression results suggest that credit risk has a positive and statistically significant effect on Tobin's Q, with a positive coefficient of 1.278545 ($p = 0.001$). This observation is indicative that companies that have less chances to be in financial distress because of this trait, according to the acquiring of Z-score, are being bid off with premiums in the market. This result supports Hypothesis 1, which postulates that proper management strategy regarding credit risk can enhance investor confidence because of financial well-being, operational safety, and reduction in the likelihood of defaulting. According to MPT, it is found that companies capable of eliminating unsystematic risk through the effective policies of good credit management are perceived to be less risky and an attractive investment target; therefore, they are perceived to have a valuation premium. This is consistent with the literature, which reported the same results, as the responsible credit management produces beneficial impacts on the attitude toward capital markets and the worth of companies (Fama and French, 1992, and Hovakimian and Li, 2011).

Conversely, beta, which is a representation of market risk, has a statistically insignificant positive coefficient of 0.1012746 ($p = 0.176$). The findings support Hypothesis 2, according to which market risk has a positive albeit insignificant effect on enterprise value. The test is not statistically significant, meaning that systematic market volatility, although applicable in the short term to returns variability, does not have a significant impact on long-term firm valuation as represented by Tobin's Q. This finding can be explained by the fact that beta only captures the fraction of risks that are important to investors and does not take into consideration firm-specific factors, including the stability of cash flows, ability to innovate, and the quality of governance. Moreover, fundamental to the firms and industry performance can tend to be the dominant factors in valuation in emerging markets such as India, compared to volatility in the market. These observations are consistent with those of Campbell and Shiller (1988) and Ang et al. (2006), who believe that beta is not an adequate measure of risk that is relevant to valuation. The interpretation is also enhanced by the control variables. The size of the firm has a positive significant coefficient (0.2316587, $p = 0.004$), which indicates that an enterprise valuation is likely to increase with firm size because of the perceived stability, resource strengths, and scalability. There is also a very high positive correlation coefficient of leverage (2.546506, $p < 0.001$), which means that an increase in debt ratios, used effectively, can be evidence of a belief in future profitability and tax benefits, which will increase the value of firms. Enterprise capitalization, on the same note, shares a positive and significant effect (8.90e-08, $p = 0.003$), and this supports the perception that better capitalized firms are perceived positively in financial markets because of the power of its resources and its ability to grow. In order to confirm the strength of these results, the second dependent variable, PBR, was implemented in a supplemental FE regression equation. The results of the robustness test, presented in Table 11.6, provide an R^2 value of 0.6797, which implies that 67.97% of variations in PBR are accounted by the identical collection of explanatory variables. These results are similar to the ones the main model generates; credit risk has a significant and positive impact on PBR, but the relationship between market risk and the latter is yet to be determined as statistically significant. This is indeed the fact that similarities between the two measures of valuation, Tobin's Q and the PBR, provides the affirmation that credit risk is a key determinant of enterprise valuation, and the effect of market risk is not so high.

Financial risk management is a key corporate strategy and has a major impact on the valuation of firms. In the dynamic and complicated financial conditions the firms operate in, it is important to understand the differentiated impacts of different financial risks. Financial risk is generally broken down into credit risk and market risk, and both of them have an impact on the financial health of firms, investor confidence, and long-term value. The present study considers the role of financial risk in the valuation of nonfinancial firms that are listed in the BSE 500 index between the years 2017 and 2023 with the help of a panel data comprising 1,988 firm-years of 284 firms. This analysis is done using Tobin's Q and the PBR as the complementary valuation measures to determine the role of credit and market risks in determining the value of a firm. It has been empirically proved that credit risk measured by the Altman *Z*-score positively influences enterprise value and has a statistically significant positive effect on enterprise value, whereby firms with strong credit fundamentals and less financial distress are more likely to experience high market valuations. This helps confirm that the MPT is correct and consistent with the findings of prior empirical research, showing the importance of credit risk assessment, monitoring, and mitigation as a part of its strategy. The outcome highlights the idea that good credit risk management leads to increased investor confidence, financial strength, and value of firms, among other things. On the other hand, market risk, which is indicated by the beta coefficient, does not show a statistically significant association with firm value in both model specifications. This indicates that market changes can have a short-term effect on the volatility of the earnings, but it does not always have a direct effect on long-term valuation parameters such as Tobin's Q and the PBR. This result is consistent with the literature like Campbell and Shiller (1988) or Ang et al. (2006), which doubts the adequacy of beta in measuring the entire range of risk of valuation, especially in emerging markets where specific firm-related and operational risks can be dominant.

The study has a number of implications from a managerial perspective. First, firms should design better credit risk management mechanisms in order to improve their valuation. Strict credit screening, real-time monitoring, and stress testing of exposures can not only stabilize the operations, but it also indicates financial prudence to the capital markets. Second, market risk could not be identified as a major determinant of firm value; however, it still cannot be neglected. To be resilient to macroeconomic shocks, managers are advised to have diversified portfolios, hedge against market risks, and incorporate market risk issues into larger enterprise risk management (ERM) frameworks. The study will be useful in the literature of financial risk management as it provides empirical data on how credit and market risks have dissimilar impacts on firm valuation based on two commonly accepted measures of value. Strongness tests that are performed with the help of the PBR contribute to enhancing the trustworthiness of the results and the validity of the consistency of the results among the various firm value proxies. Nevertheless, this study does not lack limitations. First, beta as a measure of market risk might be inadequate in measuring more complex and firm-specific aspects of risk. Beta mainly captures systematic volatility in the market, and it can be insufficient to capture idiosyncratic risks, positions taken, industry effects, etc., which affect valuation. This limitation may be overcome in future studies by the inclusion of multifactor models or alternate risk tiers like idiosyncratic volatility or downside beta. Second, there is sampling of only full-filed firms' data and left out financial institutions, which can lead to survivorship bias and make it difficult to generalize the findings. Also, the use of publicly tradable companies may fail to reflect the risk dynamics in privately owned firms. Future studies must seek to expand the research framework so as to investigate how other categories of financial and operational risks, e.g., liquidity risks, operational failure risks, or risks caused by governance issues, affect the value of a firm. They could also add to the existing knowledge by comparing risk profiles in different countries or specialized sectors.

11.8 ACKNOWLEDGMENT

We would like to express our gratitude to Centurion University of Technology and Management and KIIT Deemed to be University, which provided assistance or resources for the study. Their invaluable support and guidance were crucial in completing this research.

11.9 FUNDING

No funding.

11.10 CONFLICT OF INTEREST

The authors declare that there are no conflicts of interest related to this study.

11.11 AUTHOR CONTRIBUTIONS

- **Ashutosh Parhi:** Conceived the study, performed the analysis, and wrote the manuscript.
- **Kishor Chandra Meher:** Contributed through conceptual inputs and valuable review comments that enhanced the overall quality of the manuscript.
- **Rabi Narayan Subudhi:** Contributed to the research design and provided critical revisions to the manuscript.
- **Saroj Kumar Routray:** Assisted in data collection and provided analytical support.

11.12 ETHICS APPROVAL

This study was conducted following ethical standards. Ethics approval was obtained from KIIT Deemed to be University. All participants provided informed consent before participation in the study (if applicable).

11.13 DATA AVAILABILITY

The data supporting the findings of this study are available from the corresponding author upon reasonable request.

11.14 ABBREVIATIONS

- **Tobin's Q (TQ):** A ratio used to measure a firm's market value relative to its asset value.
- **Price-to-Book Ratio (PBR):** A valuation ratio comparing a firm's market price to its book value.
- **Modern Portfolio Theory (MPT):** A framework for portfolio management and risk optimization.

- **Fixed Effects (FE)**: A statistical model used to analyze panel data.
- **Random Effects (RE)**: An alternative to FE in panel data regression.

REFERENCES

Acharya, V. V., Amihud, Y., & Bharath, S. T. (2013). Liquidity risk of corporate bond returns: Conditional approach. *Journal of Financial Economics*, 110(2), 358–386.

Ali, M., Akbar, S., & Ormrod, P. (2018). Financial risk and firm value: Evidence from non-financial UK firms. *International Review of Financial Analysis*, 58, 21–33.

Almeida, H., & Philippon, T. (2007). The risk-adjusted cost of financial distress. *Journal of Finance*, 62(6), 2557–2586.

Altman, E. I. (1968). Financial ratios, discriminant analysis and the prediction of corporate bankruptcy. *The Journal of Finance*, 23(4), 589–609.

Amihud, Y., & Mendelson, H. (1986). Asset pricing and the bid-ask spread. *Journal of Financial Economics*, 17(2), 223–249.

Ang, A., Hodrick, R. J., Xing, Y., & Zhang, X. (2006). The cross-section of volatility and expected returns. *The Journal of Finance*, 61(1), 259–299.

Banz, R. W. (1981). The relationship between return and market value of common stocks. *Journal of Financial Economics*, 9(1), 3–18.

Beasley, M. S., Clune, R., & Hermanson, D. R. (2008). The impact of enterprise risk management on the internal audit function. *Journal of Information Systems*, 22(1), 65–86.

Bodie, Z., Kane, A., & Marcus, A. J. (2014). *Investments* (10th ed.). McGraw-Hill Education, New York.

Campbell, J. Y., & Shiller, R. J. (1988). The dividend-price ratio and expectations of future dividends and discount factors. *The Review of Financial Studies*, 1(3), 195–228.

Campbell, J. Y., & Vuolteenaho, T. (2004). Bad beta, good beta. *The American Economic Review*, 94(5), 1249–1275.

Chien, Y. W., & Chen, H. L. (2021). The effect of credit risk on firm value post-financial crisis. *International Journal of Financial Studies*, 9(4), 56.

Cruz, J. M. (2013). Mitigating global supply chain risks through corporate social responsibility. *International Journal of Production Research*, 51(13), 3995–4010.

Daud, S., Ahmad, R., & Khalid, A. R. (2011). The impact of market risk on bank performance: Evidence from Malaysia. *International Journal of Business and Social Science*, 2(20), 248–259.

Debby, C., Chan, K., & Lee, H. (2014). Corporate governance and firm valuation in Asian emerging markets. *Journal of Business Ethics*, 122(4), 579–598.

Demsetz, H., & Villalonga, B. (2001). Ownership structure and corporate performance. *Journal of Corporate Finance*, 7(3), 209–233.

Demirgüç-Kunt, A., & Huizinga, H. (2000). Financial structure and bank profitability. *World Bank Economic Review*, 13(2), 379–408.

Elton, E. J., & Green, T. C. (2003). Tax and liquidity effects in pricing government bonds. *Journal of Finance*, 58(2), 747–774.

Erasmus, P. D. (2008). *Evaluating value-based financial performance measures* (Doctoral dissertation, Stellenbosch: University of Stellenbosch).

Erin, O., Olayinka, O. B., & Ayodotun, I. O. (2018). Financial risk and firm value: A study of non-financial firms listed on the Nigerian Stock Exchange. *Journal of Business and Economic Development*, 3(2), 38–49.

Fama, E. F., & French, K. R. (2015). A five-factor asset pricing model. *Journal of Financial Economics*, 116(1), 1–22.

Fama, E. F., & French, K. R. (1992). The cross-section of expected stock returns. *The Journal of Finance*, 47(2), 427–465.

Ferson, W. E., & Harvey, C. R. (1991). The variation of economic risk premiums. *Journal of Political Economy*, 99(2), 385–415.

Ganesh, R., Kumar, M., & Rajesh, S. (2013). The effects of financial leverage on firm value: Evidence from non-financial firms. *Journal of Business and Economic Policy*, 1(2), 95–106.

Ghani, M., & Nayan, S. (2016). Financial risk management and firm performance: Evidence from Malaysia. *Journal of Management Sciences*, 10(3), 1–12.

Ghosh, S., & Nanda, R. (2022). The impact of sovereign debt crisis on firm value and credit risk management in Europe. *European Financial Management*, 28(3), 345–374.

Giesecke, K., Longstaff, F. A., Schaefer, S., & Strebulaev, I. A. (2011). Corporate bond default risk: A 150-year perspective. *Journal of Financial Economics*, 102(2), 233–250.

Gordon, L. A., Loeb, M. P., & Tseng, C. Y. (2009). Enterprise risk management and firm performance: A contingency perspective. *Journal of Accounting and Public Policy*, 28(4), 301–327.

Hausman, J. A. (1978). Specification tests in econometrics. *Econometrica*, 46(6), 1251–1271.

Heston, S. L., Rouwenhorst, K. G., & Wessels, R. E. (1995). The role of beta and size in the cross-section of European stock returns. *European Financial Management*, 1(2), 139–159.

Holton, C. (2009). Identifying disgruntled employee systems fraud risk through text mining: A simple solution for a multi-billion-dollar problem. *Decision Support Systems*, 46(4), 853–864.

Hovakimian, A., & Li, G. (2011). In search of conclusive evidence: How to compare credit ratings and credit risk. *Journal of Financial Economics*, 100(2), 282–297.

Hubbard, D. W. (1998). *How to measure anything: Finding the value of "intangibles" in business*. Wiley, Hoboken.

Imbierowicz, B., & Rauch, C. (2014). The relationship between liquidity risk and credit risk in banks. *Journal of Banking & Finance*, 40, 242–256.

Jegadeesh, N., & Titman, S. (2019). Cross-sectional and time-series determinants of momentum returns. *Review of Financial Studies*, 32(5), 1604–1647.

Kothari, S. P., Shu, S., & Wysocki, P. D. (2009). Do managers withhold bad news? *Journal of Accounting Research*, 47(1), 241–276.

Li, J., & Zhang, Y. (2023). Credit risk and firm value in the wake of COVID-19: The role of risk management. *Journal of Financial Stability*, 65, 101054.

Lintner, J. (1965). The valuation of risk assets and the selection of risky investments in stock portfolios and capital budgets. *The Review of Economics and Statistics*, 47(1), 13–37.

Litzenberger, R. H., & Ramaswamy, K. (1979). The effect of personal taxes and dividends on capital asset prices: Theory and empirical evidence. *Journal of Financial Economics*, 7(2), 163–195.

Liu, X., & Wang, Y. (2015). Credit risk and firm value in China: The role of risk management practices. *Asia-Pacific Financial Markets*, 22(3), 219–234.

Markowitz, H. (1952). Portfolio selection. *The Journal of Finance*, 7(1), 77–91.

Matayo, O., & Muturi, W. (2018). The effect of credit risk management on the financial performance of savings and credit cooperative societies in Nairobi County, Kenya. *International Journal of Economics, Commerce and Management*, 6(5), 788–806.

Merton, R. C. (1974). On the pricing of corporate debt: The risk structure of interest rates. *The Journal of Finance*, 29(2), 449–470.

Nguyen, T. T., & Tran, Q. T. (2023). FinTech innovation and credit risk management: Implications for firm value. *Journal of Financial Innovation*, 10(2), 155–172.

Ohlson, J. A. (1980). Financial ratios and the probabilistic prediction of bankruptcy. *Journal of Accounting Research*, 18(1), 109–131.

Pagach, D., & Warr, R. (2011). The characteristics of firms that hire chief risk officers. *Journal of Risk and Insurance*, 78(1), 185–211.

Rajan, R. G., & Zingales, L. (1995). What do we know about capital structure? Some evidence from international data. *The Journal of Finance*, 50(5), 1421–1460.

Rossi, S., Schwaiger, M. S., & Winkler, G. (2013). Credit risk transfer activities and systemic risk: How banks became less risky individually but posed greater risks to the financial system at the same time. *Journal of Financial Stability*, 9(3), 399–414.

Sharpe, W. F. (1964). Capital asset prices: A theory of market equilibrium under conditions of risk. *The Journal of Finance*, 19(3), 425–442.

Siringi, S. M., & Obange, N. (2021). Credit risk management and financial performance of commercial banks in Kenya. *Global Journal of Management and Business Research*, 21(2), 16–24.

Sisay, A. (2017). Credit risk management and profitability of selected commercial banks in Ethiopia. *Research Journal of Finance and Accounting*, 8(7), 95–103.

Titman, S., & Wessels, R. (1988). The determinants of capital structure choice. *The Journal of Finance*, 43(1), 1–19.

Varotto, S. (2011). Liquidity risk, credit risk, market risk, and bank capital. *International Journal of Managerial Finance*, 7(2), 134–152.

Walker, C. F., & Black, R. E. (2004). Zinc and the risk for infectious disease. *Annu. Rev. Nutr.*, 24(1), 255–275.

Zhang, R., & Zhang, Y. (2011). Credit risk management and firm value in emerging markets. *Emerging Markets Review*, 12(2), 139–152.

Chapter 12

Bilevel Programming Application in Cardinality-Constrained Portfolio Model

Bhuvnesh Khatana and Geetanjali Panda

12.1 INTRODUCTION

Markowitz's model, commonly known as the mean–variance portfolio optimization model, obtains the optimal portfolio by considering the expected return and the covariance of the portfolio's returns as inputs. Apart from covariance-based risk, several other risk measures are known in the literature, such as value at risk (Morgan [1], Lwin et al. [2]), conditional value at risk (Rockafellar and Uryasev [3]), and mean absolute deviation (Konno and Koshizuka [4]). Markowitz's model is efficient and widely used, but it fails to address real-world trading scenarios in its original form. A number of real-world trading constraints need to be included with the model, such as cardinality constraints, transaction costs, and bounding constraints. Each purchase or sale of assets incurs transaction costs on the investor, which significantly impact the overall investment, especially in large-scale portfolios. These costs typically include broker fees, bank charges, and taxes associated with asset trading. Konno and Wijayanayake [5] demonstrated that transaction costs are relatively high for small investments and show concave behavior initially and change to convex behavior beyond a certain threshold. Lobo et al. [6] showed that transaction costs in the portfolio selection process can be modeled using fixed and linear functions. Investors often restrict the size of the portfolio by controlling the number of assets in the portfolio, which is achieved through the cardinality constraint (see Yen and Yen [7], Hooshmand et al. [8]). Portfolios with cardinality constraints in the models are significant, as they reduce transaction costs and simplify portfolio management. Another common constraint bounds the proportion of capital of individual assets to ensure realistic asset holdings. In the optimal portfolio, the weights of investment in different assets are associated with minimum risk, maximum return of investment, or a compromise between both. The performance of the portfolio is assessed to determine the quality of the optimal solution. Numerous performance measures (see Marhfor [9]) are used for this purpose, such as the information ratio, Sortino ratio, Sharpe ratio (SR), and Treynor ratio. This chapter uses the SR as a performance measure

 DOI: 10.1201/9781003642091-12

of the optimal portfolio. Sharpe [10] evaluates risk-adjusted returns and is calculated as the value of the expected return on the unit risk.

There are several hierarchical situations in the trading market, where a financial strategy is associated at each level to optimize several financial factors while selecting an efficient portfolio. These decision-making situations can be modeled as bilevel optimization problems. These models are formulated in different scenarios at both levels. In the hierarchical model by Benita et al. [11], the investor determines optimal capital for a decentralized global investment strategy through an intermediary model of optimizing the return of risky assets. Leal et al. [12] and González-Díaz et al. [13] proposed bilevel portfolio models in which the broker–dealer system establishes a hierarchical structure between them. The broker acts as a leader and sets transaction costs with the portfolio, while the investor acts as a follower and minimizes the tail risk. Bayat et al. [14] considered the risk budgeting problem using a bilevel formulation in which the upper level is associated with risk budgets and optimizes expected return, whereas the risk budgeting portfolio is optimized in the lower level. Kobayashi et al. [15] considered a mean–risk portfolio model with conditional value at risk, including the cardinality constraint, which is then reformulated as a bilevel model to deal with the complexity of cardinality. Cesarone et al. [16] modeled a problem in which financial service providers construct ESG (environmental, social, and governance)-oriented portfolios to optimize the firm's overall ESG impact. Multiple account holders acting as followers determine optimal portfolios by considering risk, return, transaction costs, and ESG scores in the model while maintaining a Nash equilibrium among themselves. Chen et al. [17] and Jing et al. [18] estimated parameters in a mean–variance model, such as expected return, cardinality, and weight bounds, while maximizing the SR at the upper level.

In this chapter, we consider a bilevel portfolio optimization model, which is different from the above models. The upper level of the proposed model maximizes the SR, and the lower level minimizes the covariance risk of the portfolio, with the expected return and cardinality of the portfolio being parameterized. The upper-level problem is not necessarily a concave problem, as the SR is not a concave function in general. The lower-level problem is formulated as a nonconvex quadratic optimization problem due to the presence of the cardinality constraint. The cardinality constraint introduces integer variables in the lower level of the bilevel model. This type of bilevel problem cannot be solved directly by classical optimization techniques. In the present work, the concept of the branch-and-bound technique is used at the lower level to solve the bilevel model. The methodology is summarized as an algorithm, and it is justified that the algorithm stops at an optimal portfolio.

To facilitate a portfolio model, investors collect historical data from the stock market for a time horizon with regular intervals. The common assumption on asset returns is that they are normally distributed and uncorrelated, which is not necessarily true for real-world stock data. This causes potential inaccuracies in the estimation of inputs such as mean and variance, leading to suboptimal solutions in portfolio models. As pointed out in the work of Best and Grauer [19], this uncertainty primarily arises from the unknown probability distribution of the input parameters and the model's high sensitivity to variations with respect to these parameters. In practice, the true values of these parameters cannot be determined with certainty. Broadie [20] indicated that using a limited historical dataset for mean–variance analysis can introduce estimation errors of the values in the input parameters, while for long-term datasets, it may lead to non-stationarity of the parameters in the data. To address these challenges, various well-known techniques are used in the literature, such as the Bayesian statistics technique (Jorion [21], Bauder et al. [22]), the stochastic optimization technique (Fernholz and Fernholz [23]), the robust optimization technique (Moon and Yao [24]), simulation techniques (Jorion [25]), and fuzzy portfolio optimization (Fang et al. [26]). In the proposed methodology, we use the method employed by Jorion [25], which uses a simulation technique to obtain the optimal portfolio by handling the estimation problem of the mean and variance of the return of the assets.

The present work is divided into several sections. In Section 12.2, problem formulation and the associated model are proposed. Section 12.3 outlines the methodology to solve the proposed model. In Section 12.4, we address the implementation of the methodology in financial data from Bombay Stock Exchange (BSE), India.

12.2 MODEL FORMULATION

A bilevel optimization problem is an optimization model of hierarchical structure in which another parametric optimization problem appears as a constraint nested with the main optimization problem. The bilevel optimization problem is closely related to the Stackelberg game in game theory. In this game, if the mode of play is cooperative and the follower selects the strategy that is favorable to the leader, the bilevel formulation is categorized as optimistic. The upper level or leader's problem is

$$\min_{\chi} \Theta_U(\chi,\gamma) \text{ s.t. } \xi_U(\chi,\gamma) \le 0,$$

where $\Theta_U : \mathbb{R}^n \times \mathbb{R}^m \to \mathbb{R}$ and $\xi_U : \mathbb{R}^n \times \mathbb{R}^m \to \mathbb{R}^p$.

The vector γ in the upper-level problem is determined by solving the following lower-level or follower's problem:

$$\min_{\gamma} \Theta_L(\chi,\gamma) \text{ s.t. } \xi_L(\chi,\gamma) \le 0,$$

where $\Theta_L : \mathbb{R}^n \times \mathbb{R}^m \to \mathbb{R}$ and $\xi_L : \mathbb{R}^n \times \mathbb{R}^m \to \mathbb{R}^q$. A general optimistic bilevel optimization problem with inequality constraints is

$$
(P): \begin{array}{ll} \min\limits_{\chi,\gamma} & \Theta_U(\chi,\gamma) \\ \text{s.t.} & \xi_U(\chi,\gamma) \le 0, \\ & \gamma \in \underset{\gamma}{\operatorname{argmin}} \left\{ \Theta_L(\chi,\gamma) \text{ s.t. } \xi_L(\chi,\gamma) \le 0 \right\}. \end{array}
$$

A detailed discussion on the various important aspects of general bilevel optimization problems, including the optimality conditions and classical approaches of solution methodology, is presented in the monograph by Dempe [27]. A common approach to solving a bilevel optimization problem is to reformulate the original model into a single-level optimization problem. This is mainly achieved by using the optimality conditions on the lower-level problem of the bilevel formulation. In the case when the lower-level problem is convex and regular, Karush–Kuhn–Tucker (KKT) optimality conditions of the lower-level problem can be incorporated at the upper level. This approach introduces the dual variables along with the complementarity constraints associated with the KKT optimality conditions. Since the lower-level problem is the convex programming problem, the resulting single-level problem can be characterized as Mathematical Programs with Equilibrium Constraints (MPECs). One difficulty in this transformation is that the resulting MPEC is not equivalent to the original bilevel optimization problem in the sense of the optimal solutions, as pointed out in the study by Dempe and Dutta [28]. However, if the linear independence constraint qualification holds for the lower-level problem, then the KKT dual vectors are unique, and both problems are proved to be equivalent in terms of solution. In fact, under these restrictions, MPEC approaches can be adopted to solve the bilevel optimization problem. We explain below about the MPEC formulation of the bilevel model P.

The Lagrange function for the lower-level problem of P is $L(\chi,\gamma,\eta) = \Theta_L(\chi,\gamma) + \bullet^T \xi_L(\chi,\gamma)$, where $\eta \in \mathbb{R}^q$ is the dual vector associated with the constraint function ξ_L. The KKT optimality conditions for the lower-level problem are given in the set Ω as follows:

$$
\Omega = \left\{ (\chi,\gamma,\eta) : \nabla_\gamma L(\chi,\gamma,\eta) = 0, \xi_L(\chi,\gamma) \le 0, \eta \ge 0, \eta^T \xi_L(\chi,\gamma) = 0 \right\}
$$

The corresponding MPEC formulation is

$$
\underset{\chi,\gamma,\eta}{\text{Min}}\, \Theta_U(\chi,\gamma)
$$

$$
\text{s.t. } \xi_U(\chi,\gamma) \le 0, (\chi,\gamma,\eta) \in \Omega
$$

Next, we formulate the portfolio optimization model as a bilevel optimization problem. The following notations are used to explain the formulation of the model.

Notations:

- A_i : i^{th} asset of the portfolio $P = (A_1, A_2, \cdots, A_n)$.
- T : Total number of observations for the collected data.
- r_{ij} : Return of asset A_i at time t_j.
- σ_{ij} : Covariance between the returns of assets A_i and A_j.
- μ_i : Expected return of asset A_i.
- $\mu := (\mu_1, \mu_2, \cdots, \mu_n)^T$.
- $w := (w_1, w_2, \cdots, w_n)^T, \sum_{i=1}^{n} w_i = 1$, where w_i is the proportion of the invested capital in A_i.
- $\|w\|_0$: Number of nonzero entries in vector w.
- $\rho := \mu^T w$ Expected return of the portfolio or investment.
- Σ : Covariance matrix of the return of the assets.
- $u := \begin{pmatrix} u_1 \\ u_2 \end{pmatrix} \in \mathbb{R}^2$, and $v := \begin{pmatrix} v_a \\ v_b \end{pmatrix} \in \mathbb{R}^{n+n}$ are the Lagrange multiplier vectors associated with the equality and inequality constraints of the lower-level problem of the bilevel model, respectively.

The cardinality constraint provides a choice to the investor to select the number of assets to be included in the portfolio from a given set of assets. The cardinality constraint is expressed as $\|w\|_0 \leq \beta$, which implies that the nonzero weights will be assigned to most β assets (out of a total of n assets) in the optimal portfolio, and the weight for the remaining $n - \beta$ assets is set to be zero. The value of β is typically determined by the investor before determining the optimal portfolio. Jing et al. [18] determined empirically that the different values of β and expected return ρ can impact the portfolio's performance. Hence, it may be tempting for the investor to search for the appropriate pair (ρ, β) that corresponds to the maximum performance of the portfolio.

On the other hand, if there are fewer assets in the optimal portfolio, then it may be helpful to reduce the estimation error and transaction costs associated with the assets. The estimation error increases as the size of the portfolio grows for a fixed sample size. Lobo et al. [6] showed that there are fixed transaction costs associated with the trading of the assets. In addition to this, transaction costs can be represented by linear functions, which are proportional to the amount of the asset bought or sold in the portfolio. The transaction cost for the ith asset is given by $\phi_i(w_i) = \alpha_i^+ w_i$ if $w_i \geq 0$, *and* $\phi_i(w_i) = \alpha_i^- w_i$ if $w_i < 0$, where α_i^+ and α_i^- are the proportional cost rates for buying and selling, respectively. The total transaction cost of the portfolio is the sum of individual transaction costs over all assets, computed by $\sum_{i=1}^{n} \phi_i(w_i)$. Hence, setting $w_i = 0$ for some of the assets reduces the transaction costs.

This chapter structures the SR and the risk factor of the portfolio into two levels, which can be treated as a bilevel portfolio optimization model. The upper level maximizes the SR of the portfolio, whereas the lower level minimizes the risk for certain returns and cardinality of the assets. The SR of the portfolio is defined as the ratio of the excess return of the portfolio and the risk and is formulated mathematically as $\frac{\mu^T w - r_f}{\sqrt{w^T \Sigma \mathrm{w}}}$, where r_f is the rate of return of a risk-free asset and $\mu^T w = \rho$ is the expected return.

The following cardinality-constrained bilevel portfolio model (BPM) of a single-period investment portfolio is considered. The objective of the investor for the formulation of such a model is to identify a suitable set of assets from n assets in the market with optimal allocation of the investment in this set of assets and expected return of the investment that achieves the maximum SR.

$$(BPM): \quad \max_{\rho \geq \alpha, \beta} \frac{\rho - r_f}{\sqrt{w^T \Sigma w}}, \tag{12.1}$$

where $w \in argmin\, w^T \Sigma w$ s.t. $e^T w = 1, \mu^T w = \rho, a \leq w \leq b, \|w\|_0 \leq \beta$.

The upper level of the portfolio optimization model is

$$(UP): \quad \max_{\rho, \beta} \frac{\rho - r_f}{\sqrt{w^T \Sigma w}} \text{ s.t. } \rho \geq \alpha,$$

where w is determined through a lower-level problem, which is

$$\left(LP_{(\rho,\beta)}\right): \quad \min_{w} w^T \Sigma w \text{ s.t. } e^T w = 1, \mu^T w = \rho, a \leq w \leq b, \|w\|_0 \leq \beta, \tag{12.2}$$

where $a, b \in \mathbb{R}^n$ and the inequality $a \leq w \leq b$ is considered componentwise. α is a small positive number which ensures that the model provides a positive expected return on the investment, and β determines the size of the optimal portfolio. The weight bounds a and b ensure diversification in the optimal portfolio. The lower bound prevents too little investment in any asset, while the upper bound limits how much capital can be allocated to an asset for diversification.

For a given expected return ρ and cardinality β in the lower-level problem $LP_{(\rho,\beta)}$ as provided in equation (12.2), an optimal risk $w(\rho, \beta)$ can be obtained by solving $LP_{(\rho,\beta)}$, and the corresponding SR $\frac{\rho - r_f}{\sqrt{w(\rho,\beta)^T \Sigma w(\rho,\beta)}}$

can be evaluated. The upper-level problem determines the value of ρ and β that maximize the SR. The BPM identifies the optimal expected return ρ^*, cardinality, β^*, and portfolio weights w^* such that the SR is maximized and risk is minimized at the optimal expected return ρ^* and optimal cardinality β^*. This two-level formulation cannot be solved by classical optimization techniques due to the presence of cardinality constraints. We propose a methodology to deal with this difficulty in the next section.

12.3 METHODOLOGY

The main challenge in the BPM is caused by the cardinality constraint in the lower-level problem. This constraint $\|w\|_0 \leq \beta$ is inherently discrete by nature. We use the approach by Chang et al. [29] to handle the cardinality constraints located at the lower-level problem.

Denote $A := \text{diag}(a) \in \mathbb{R}^{n\times n}$ as the diagonal matrix whose diagonal entries are components $a_i, i = 1,2,\cdots,n$. Similarly denote $B := \text{diag}(b) \in \mathbb{R}^{n\times n}$. The cardinality constraint $\|w\|_0 \leq \beta$ can be restructured by

$$As \leq w \leq Bs,$$

$$\sum_{i=1}^{n} s_i \leq \beta,\ s_i \in \{0,1\}, i = 1,2,\cdots,n.$$

In that case, $LP_{(\rho,\beta)}$ model is reformulated in a mixed integer programming problem, where $s = (s_i) \in \mathbb{R}^n$ is a vector of integer variables, and $w \in \mathbb{R}^n$ are vectors of continuous variables. Hence, the BPM becomes

$$(\overline{BPM}): \quad \max_{\rho\geq\alpha,\beta} \frac{\rho - r_f}{\sqrt{w^T\Sigma w}}$$

$$\text{where } (w,s) \in \arg\min w^T\ \Sigma w$$

$$\text{s.t. } e^T w = 1, \mu^T w = \rho, As \leq w \leq Bs,$$

$$\sum_{i=1}^{n} s_i \leq \beta, s_i \in \{0,1\}, i = 1,2,\cdots,n.$$

The relaxed version of the lower-level problem is obtained by relaxing the integer constraints. A relaxed problem is solved by dropping the integer constraints, where the integer variables $s_i \in \{0,1\}$ are replaced with $0 \leq s_i \leq 1$, and s_i are treated as continuous variables for $i = 1,\cdots,n$ initially. In that case, the relaxed lower-level problem is a convex quadratic programming problem, which is easy to handle and has a unique solution for the fixed value of ρ and β.

$LP_{(\rho,\beta)}$ can be restricted in a relaxed form $LP^R_{(\rho,\beta)}$ after replacing the discrete variables $s_i \in \{0,1\}$ by continuous variables $0 \le s \le e$ as

$$LP^R_{(\rho,\beta)}: \ \min_{w,s} \ w^T \Sigma w$$

$$\text{s.t.} \ \ e^T w = 1, \mu^T w = \rho,$$

$$As - w \le 0, w - Bs \le 0,$$

$$\sum_{i=1}^{n} s_i \le \beta,$$

$$0 \le s \le e.$$

Let the Lagrange function of $LP^R_{(\rho,\beta)}$ be

$$L(\rho,\beta,w,s) := w^T \Sigma w + u_1 \left(e^T w - 1\right) + u_2 \left(\mu^T w - \rho\right)$$
$$+ v_a^T (As - w) + v_b^T (w - Bs) + t_\beta \left(\sum_{i=1}^{n} s_i - \beta \right) - t_0^T s + t_1^T (s - e),$$

where $u_1 \in \mathbb{R}, u_2 \in \mathbb{R}, v_a \in \mathbb{R}^n, v_b \in \mathbb{R}^n, t_\beta \in \mathbb{R}, t_0 \in \mathbb{R}^n, t_1 \in \mathbb{R}^n$ are the Lagrange multipliers. The KKT optimality conditions on the relaxed lower-level problem are

$$2\Sigma w + u_1 e + u_2 \mu - v_a + v_b = 0, \tag{12.3}$$

$$A^T v_a - B^T v_b - t_0 + t_1 - e t_\beta = 0, \tag{12.4}$$

$$e^T w = 1, \ \mu^T w = \rho, \ As \le w \le Bs, \tag{12.5}$$

$$\sum\nolimits_{i=1}^{n} s_i \le \beta, 0 \le s \le e, \tag{12.6}$$

$$v_a^T (As - w) = 0, v_b^T (w - Bs) = 0, \tag{12.7}$$

$$t_0^T s = 0, t_1^T (e - s) = 0, t_\beta \left(\sum\nolimits_{i=1}^{n} s_i - \beta \right) = 0, \tag{12.8}$$

$$v_a, v_b, t_0, t_1 \ge 0, \ t_\beta \ge 0. \tag{12.9}$$

Consider the relaxed form of $\overline{BPM}$ as

$$(P_R): \max F(\rho, w) \ s.t. \ \rho \ge \alpha, \ (w,s) \ solves \ LP^R_{(\rho,\beta)},$$

where $F(\rho,w):=\dfrac{\rho-r_f}{\sqrt{w^T\Sigma w}}$. Using the above KKT optimality conditions (12.3)–(12.9) for $LP^R_{(\rho,\beta)}$, the problem P_R can be restructured as

$$\max_{\rho,\beta,w,s,u,v,t} \quad F(\rho,w) \text{ s.t. } (\rho,\beta,s,u,v,t) \text{ satisfies (12.3)-(12.9)},$$

where $t=\begin{pmatrix} t_0 \\ t_1 \\ t_\beta \end{pmatrix} \in \mathbb{R}^{m+n+1}$. To facilitate the methodology, we assume that F is bounded in the relaxed constraints set (12.3)–(12.9).

Due to the presence of complementarity conditions (12.7)–(12.8), the traditional methodologies cannot provide the solution of the BPM directly. It can be solved using MPEC approaches, such as Fortuny-Amat and McCarl [30], Jiang and Ralph [31], and Kanzow and Schwartz [32]. Here, we use the Big-M method of Fortuny-Amat and McCarl [30] to solve P_R and develop a branch-and-bound method to address the integer restriction in the BPM. Some important steps of this process are explored below, and the complete process is summarized in an algorithm.

12.3.1 Branch-and-Bound Method for BPM

The proposed branch-and-bound process explores a sequence of nodes and edges, which can be visualized as a tree-like graph structure, as explained in Figure 12.1. Each node is associated with the relaxed problem. A live node in the tree is the node that is generated but has not yet been explored. The fathomed node is the node that is discarded, and further branching of that node is stopped. New nodes are generated by branching the current node, where the additional constraints in the relaxed problems are imposed. Further branching is not possible if the relaxed problem provides a feasible solution to the main problem. The optimal values of the relaxed problems are used as bounds, which set the criteria for the fathoming of the node. If this bound is not an improvement over the best-known feasible solution, further branching at that node is stopped. This organized search for the relaxed problem continues until each live node is explored. In that case, the method is terminated with the optimal solution of the main problem. This process is explained mathematically in the following manner.

- Let the *ith* node in the search tree be denoted by N_i, with the root node as N_0. Let R denote the feasible set of P_R. $R=\{(\rho,\beta,w,s,u,v,t):\rho\geq\alpha, (\rho,\beta,s,u,v,t) \text{ satisfies (12.3)-(12.9)}\}$. To initiate the process, consider $R^0=R$. Let P_{R^i} be the relaxed problem

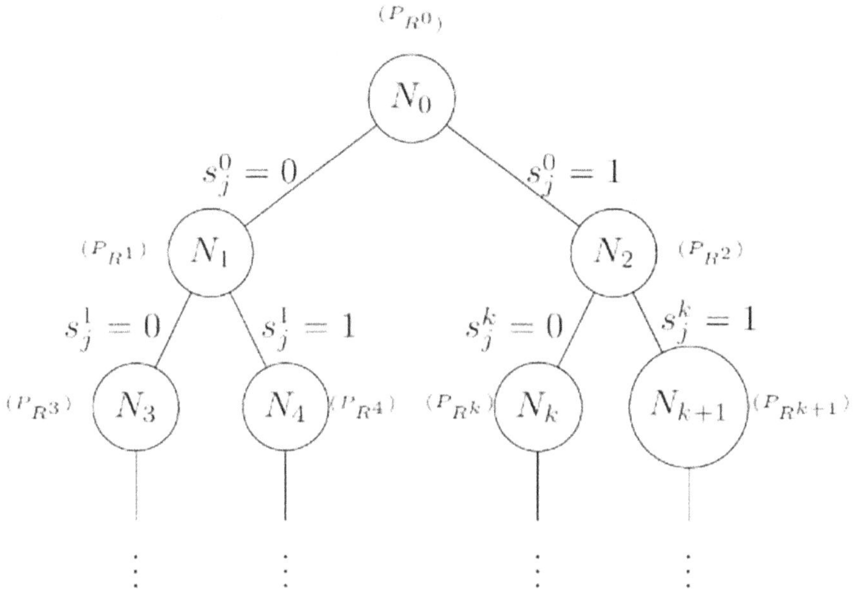

Figure 12.1 Branching in Algorithm 1

associated with the node N_i, R_i be the feasible region for P_{R^i}, and F_{R^i} be the optimal value of P_{R^i}.

$$(P_{R^i}): \max F(\rho, w) \text{ s.t. } (\rho, \beta, w, s, u, v, t) \in R^i,$$

where ρ, β are real numbers and w, s are real vectors.

Let $(\rho^i, \beta^i, w^i, s^i)$ be the optimal solution of P_{R^i}. In this solution, s^i may not be an integer vector. A non-integer component s_j^i of s^i is divided into two branches $s_j^i = 1$ and $s_j^i = 0$ at node N_i. In case of more than one non-integer component in s^i, one can select one of these non-integer components randomly. Consider the following two feasible regions corresponding to two branches $s_j^i = 0$ and $s_j^i = 1$. $R^{i+1} = \{(\rho, \beta, w, s, u, v, t) \in R^i, s_j = 0\}$ and $R^{i+2} = \{(\rho, \beta, w, s, u, v, t) \in R^i, s_j = 1\}$, where s_j in R^{i+1} and R^{i+2} corresponds to j^{th} non-integer component of s^i with $s_j^i = 0$ and $s_j^i = 1$, respectively, in the optimal solution of P_{R^i}.

- $P_{R^{i+1}}$ is solved at node N_{i+1}. The problem $P_{R^{i+2}}$ corresponds to the live node N_{i+2}, which will be explored later in the process.
- The process continues until a node is fathomed. Let $\overline{F}$ denote the best-known feasible solution, which is initially set to a large number.

Node N^i is fathomed if the problem P_{R^i} is not feasible, or $F_{R^i} \leq \overline{F}$, or components of vector s^i are all integers. Once a node is fathomed in the search tree, we propagate back to the live nodes in the tree and continue with a similar process.

Let E denote the set of live nodes. The following algorithm summarizes the above procedure.

Algorithm 1: Branch-and-Bound Method for BPM

Step 1: (Start) Begin with node $i = 0$. Set $R^0 = R$, and consider a large number $\overline{F} = \mathcal{M}$.
Step 2: (Branching) If there is no live node left, go to Step 6. Otherwise, choose a live node and go to Step 3.
Step 3: (Fathoming) Solve the relaxed problem P_{R^i} and obtain the optimal value F_{R^i} Perform the following actions sequentially.
(**3a**). If P_{R^i} is not feasible, then fathom node i and go to Step 2.
(**3b**). If $F_{R^i} \leq \overline{F}$, then fathom node i and go back to Step 2.
(**3c**). If the components of vector s^i are all integers, then fathom the node and $\overline{F} \leftarrow F_{R^i}$ and go to Step 2.
(**3d**). Otherwise, proceed to Step 5.
Step 4: (Partitioning) Choose a non-integer component s_j^i of vector s^i and create two new nodes by partitioning the current node. Go to Step 2.
Step 5: If $\overline{F} = \mathcal{M}$, then the bilevel model is infeasible. Otherwise, terminate with the current node, and the optimal value is $\overline{F}$.

In the process of fathoming the nodes, the method finds a live node by backtracking in the tree. Algorithm 1 solves a sequence of relaxed problems, P_{R^i}, by updating the relaxed sets. Using the backtracking process, Algorithm 1 explores possible nodes to increase the objective value F. The method stops when further branching is not possible and no live node exists with a larger objective value.

The following proposition proves that the solution obtained through the above algorithm is the solution to the BPM.

Proposition 3.1. Let $(\rho^*, \beta^*, w^*, s^*)$ *be the termination point of Algorithm 1. Then* (ρ^*, β^*, w^*) *is the optimal solution to BPM.*

Proof. To prove the proposition, we only need to show that w^* is the optimal solution to the $LP_{(\rho^*,\beta^*)}$ model. Suppose this is not the case, and there is an optimal point $\overline{w} \neq w^*$ of $LP_{(\rho^*,\beta^*)}$ such that $w^{*T}\Sigma w^* > \overline{w}^T \Sigma \overline{w}$. In that case, there exists an integer vector $\overline{s}$ with $\overline{s}_i \in 0,1, i = 1,2,\cdots,n$ such that $(\rho^*, \beta^*, \overline{w}, \overline{s})$ satisfies

$$e^T \overline{\mathrm{w}} = 1,\ \mu^T \overline{w} = \rho^*,\ A\overline{s} \leq \overline{w} \leq B\overline{s},\ \sum_{i=1}^{n} \overline{s}_i \leq \beta^*, \overline{s}_i \in \{0,1\},\quad i = 1,2,\cdots,n.$$

Hence, $(\rho^*, \beta^*, \bar{w})$ is a feasible point of BPM. Since $w^{*T}\Sigma w^* > \bar{w}^T\Sigma\bar{w}$, $F(\rho^*, \bar{w}) > F(\rho^*, w^*)$. This contradicts that $(\rho^*, \beta^*, w^*, s^*)$ is the terminating point of Algorithm (1). Hence, the result follows.

Example 1. For the illustration of Algorithm 1, we consider a simple example consisting of six assets, for which the covariance matrix and mean return vector are given as

$$\Sigma = \begin{bmatrix} 0.2138 & 0.1230 & 0.1024 & 0.1372 & 0.1341 & 0.1598 \\ 0.1230 & 0.2252 & 0.0789 & 0.1066 & 0.1169 & 0.1171 \\ 0.1024 & 0.0789 & 0.2561 & 0.0845 & 0.0963 & 0.0851 \\ 0.1372 & 0.1066 & 0.0845 & 0.2336 & 0.1302 & 0.1143 \\ 0.1341 & 0.1169 & 0.0963 & 0.1302 & 0.2442 & 0.1135 \\ 0.1598 & 0.1171 & 0.0851 & 0.1143 & 0.1135 & 0.2764 \end{bmatrix}$$

and $\mu = [0.00066850.00002840.0005496 - 0.0005045 - 0.00020860.0006966]$, respectively.

In the bilevel model for this example, we select the parameter values $\alpha = 10^{-5}, \bar{F} = 10^{10}, A = 10^{-3} I_{6\times6}$, and $B = I_{6\times6}$, where $I_{6\times6}$ is the identity matrix of order 6. At each node N^i, we solve the relaxed problem

$$(P_{R^i}): \quad max\ F(\rho, w)\ s.t.\ (\rho, \beta, w, s, u, v, t) \in R^i.$$

For each i, P_{R^i} is solved using the Big-M method of Fortuny-Amat and McCarl [30]. For $i = 0$,

$$R^0 = \{(\rho, \beta, w, s, u, v, t) : \rho \geq 10^{-5}, (\rho, \beta, w, s, u, v, t)\ satisfies\ (3) - (9)\}.$$

- Initially, the set of live nodes is $E = \{N_0\}$. Then, P_{R^0} is solved, and the optimal value is obtained as $F_{R^0} = 0.0485$. In the optimal solution $(\rho^0, \beta^0, w^0, s^0, u^0, v^0, t^0)$ of $P_{R^0}, s^0 = (0.6323, 0, 0.0604, 0, 0, 0.3074)^T$ is a non-integer vector. Consider the first non-integer component of s^0, which is $s_1^0 = 0.6323$. Hence, two new nodes, N_1 and N_2, are generated by adding the constraints $s_1 = 1$ and $s_1 = 0$ to the set R^0, respectively. This results in two new relaxed sets:

$$R^1 = \{(\rho, \beta, w, s, u, v, t) \in R^0 : s_1 = 1\}\ and$$

$$R^2 = \{(\rho, \beta, w, s, u, v, t) \in R^0 : s_1 = 0\},$$

corresponding to the nodes N_1 and N_2, respectively. The set of live nodes in this stage is $E = \{N_1, N_2\}$.

- Next, we solve P_{R^1} and obtain $F_{R^1} = 0.0481$. In this solution, $s^1 = (1,0,0,0,0,0.3332)^T$, where s_6^1 is a non-integer component. Hence, we branch again and generate two new nodes, N_3 and N_4, by adding the constraints $s_6 = 0$ and $s_6 = 1$ to R^1, respectively. This yields two new relaxed sets as

  $R^3 = \{(\rho,\beta,w,s,u,v,t) \in R^0 : s_1 = 1,\ s_6 = 1\}$ and

  $R^4 = \{(\rho,\beta,w,s,u,v,t) \in R^0 : s_1 = 1,\ s_6 = 0\}$, corresponding to nodes N_3 and N_4, respectively. At this stage, $E = \{N_2, N_3, N_4\}$.

- At node N_3, we solve P_{R^3} and obtain $F_{R^3} = 0.0481$. In the solution $s_3 = (1,0,0,0,0,1)^T$, all the components are integers. Hence, we set $\overline{F} \leftarrow F_{R^3} = 0.0481$, fathom this node, and proceed to explore another live node in the tree. The set of live nodes is $E = \{N_2, N_4\}$ at this stage.
- At node N_4, we solve P_{R^4} and obtain $F_{R^4} = 0.0464$. Since $F_{R^4} < \overline{F}$, we fathom node N_4. Here, $E = \{N_2\}$. Node N_2 is explored in the next stage.
- At node N_2, we solve P_{R^2} and obtain $F_{R^2} = 0.0416$. Since $F_{R^2} < \overline{F}$, we also fathom node N_2. At the last stage, the set of live nodes is $T = \varnothing$.

  We conclude with $\overline{F} = F_{R^3} = 0.0481$ as the optimal value of the bilevel portfolio problem, and at this stage, $s^3 = (1,0,0,0,0,1)^T$. The corresponding optimal investment strategy is

  $w^3 = (0.6662,0,0,0,0,0.3338)^T$.

12.4 NUMERICAL EXPERIMENT WITH FINANCIAL DATA

Consider an investor is interested in trading n risky assets, A_i, where $i = 1,2,\ldots,n$. Let $\boldsymbol{r}_i$ represent the return of asset A_i, which is a random variable following a normal distribution $N(\mu,\sigma)$. Let σ_{ik} denote the covariance of returns between assets A_i and A_k, and r_{ij} represent the return of asset A_i at time t_j, where $j = 1,2,\ldots,T$. The expected return of asset A_i is $\mu_i := \frac{1}{T}\sum_{j=1}^{T} r_{ij}$ and $\sigma_{ik} := \frac{1}{T-1}\sum_{j=1}^{T}(r_{ij} - \mu_i)(r_{kj} - \mu_k)$.

The objective is to determine a list of assets out of n assets using Algorithm 1, which constitutes the optimal portfolio providing the maximum SR. All the numerical experiments are conducted on a Windows 10 PC using MATLAB (2023b) code for Algorithm 1. Without loss of generality, $r_f = 0$ can be considered. It is important to disable short-selling in the model to measure the effect of cardinality on the SR. In the case of short-selling, we have obtained optimal cardinality equal to the total number of assets in our numerical experiments. To solve P_{R_i} at each level in the tree, we use the Big-M method to determine the solution.

To maintain diversity in the portfolio and capture various stock trends, we selected stocks from different categories, including banking, information technology, automobile, and energy sectors, as listed in Table 12.1. For the purpose of a numerical experiment, we collected the following set of 50 stocks from BSE, India, for the duration from September 16, 2021, to July 31, 2024. The data was collected over a period of 3 years to reflect the volatility in stock prices. The historical data of the price of the assets was collected from BSE, India (https://www.bseindia.com/markets). In practice, the raw data from the stock market has outliers, which can affect the results. To identify the outliers, we used the interquartile range (IQR) method, which is based on the first quartile (Q_1) and the third quartile (Q_3) of the mean return. First, the lower and the upper bounds of the returns for each asset were calculated as $Q_1 - 1.5 \times (Q_3 - Q_1)$, and $Q_3 + 1.5 \times (Q_3 - Q_1)$, respectively. Returns falling outside these bounds were classified as outliers. Instead of removing them, the five-point moving average technique was applied to smooth these outliers' values. Furthermore, to retain the overall structure of the data while reducing fluctuations, an exponential moving average (EMA) smoother with a span of 5 previous trading days was applied to the asset price matrix.

Various types of performance measures are used to assess the quality of a portfolio, such as the Information ratio, Sortino ratio, and SR, as discussed in the Introduction section. This work has focused on the SR as the

Table 12.1 List of Stocks Used in the Numerical Experiments.

Category	*Stocks*
Information technology	Wipro, Tech Mahindra, Tata Consultancy Services, Sonata Software, Mphasis, LTIMindtree, Infosys, HCL Technologies, Cyient, Coforge
Banking	SBI, Kotak Mahindra, Yes Bank, Bandhan Bank, PNB, ICICI, HDFC, IndusInd, Bajaj Finance, Axis Bank
FMCG	Tata Consumer Products, Hindustan Unilever, Nestle India, Marico, ITC, Godrej Consumer Products, Emami, Dabur, Colgate-Palmolive, Industries
Automobile	TVS Motor Company, Tata Motors, Samvardhana Motherson, Maruti Suzuki, Mahindra & Mahindra, Hero MotoCorp, Force Motors, Eicher Motors, Bajaj Auto, Ashok Leyland
Energy	Tata Power, Reliance Industries, Petronet LNG, ONGC, NTPC, GAIL (India), Indian Oil Corporation, Coal India, Bharat Petroleum, Adani Power

performance measure. The SR is computed as $SR = \frac{\rho - r_f}{w^T Cw}$, where r_f is the risk-free return, and w is the optimal investment strategy for the given portfolio. This value shows the excess expected return over the risk-free return r_f on the unit risk. One can also adopt different performance metrics depending on the specific objective of the investor and model formulation. The information ratio accounts for a benchmark in its formula and is computed as $\frac{\rho - \rho_I}{w^T \bar{C} w}$, where ρ_I denotes the expected return of the benchmark portfolio, and $\bar{C}$ refers to the covariance matrix for the return $r - \rho_I$. One can also compute the beta value for the portfolio to observe the sensitivity of the asset's prices compared to the market return. It is computed as $\frac{Cov(r_i, r_m)}{\sigma^2(r_i)}$, where $Cov(r_i, r_m)$ is the covariance between the returns of assets A_i and market return r_m . The beta value for the portfolio is given as the weighted sum of the individual betas of the ith asset as $\sum_{i=1}^{n} w_i (beta)^i$.

The negative SR in the optimal portfolio indicates that it has underperformed relative to the risk-free asset. Moreover, a positive and higher SR relative to some benchmark portfolio suggests that better returns are achieved for the same level of risk. In financial markets, assets with lower volatility show a higher SR.

For the purpose of the numerical experiments, data is divided into two parts, namely, training data (in-sample data) and testing data (out-of-sample data). A simulation technique is employed on the in-sample data, and the optimal portfolio is estimated using this dataset. Out-of-sample data is used to test the quality of the results obtained from the in-sample period.

In-sample period: From September 16, 2021, to June 19, 2024.

Out-of-sample period: From June 20, 2024, to July 31, 2024.

Using the built-in code *datasample* of MATLAB, we obtained samples of the in-sample dataset. Let the mean and covariance matrix of the sampled data be denoted as μ^{in} and C^{in} , respectively. The bilevel model BPM is then solved by replacing μ by μ^{in} and the matrix Σ by C^{in} . Let (ρ^*, β^*, w^*) be the optimal solution of this problem. Then, the in-sample SR is calculated as $\frac{\rho^*}{\sqrt{w^{*T} C^{in} w^*}}$. Next, let C^{out} represent the covariance matrix of the out-of-sample data. Using this, the out-of-sample SR is calculated using the formula $\frac{\rho^*}{\sqrt{w^{*T} C^{\text{out}} w^*}}$. The result obtained from the out-of-sample data represents one observation of the optimal portfolio. The sampling process is repeated multiple times (in our case, 200) to obtain many observations in the optimal portfolio. It is said that the model performs well on the out-of-sample data (testing period), if the difference between the results of in-sample and

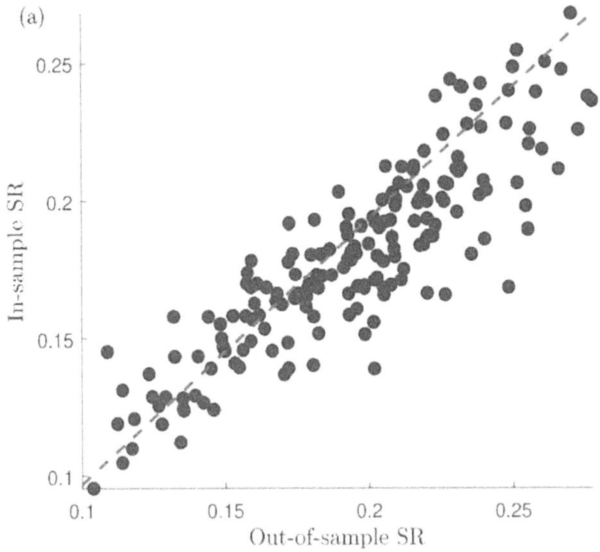

Figure 12.2 In-sample SR and out-of-sample SR of 200 samples.

out-of-sample data is smaller. In Figure 12.2, a scatter plot between the in-sample SR and the out-sample SR is shown, which is obtained for different samples. To visualize the gap between the results, a straight line $y = x$ is drawn. The figure indicates that the in-sample outcomes closely approximate the out-of-sample results, with a root-mean-square error of 0.0247.

The results obtained in the sampling process are plotted in Figure 12.3, and Table 12.2 shows the mean and median values of the optimal portfolio, obtained corresponding to the sample data of in-sample data. A simple observation is that the expected return, in-sample SR, and out-of-sample SR show a near-normal distribution around their mean values. Hence, the mean value provides a better estimate of the optimal portfolio.

In another experiment, we considered the data for the in-sample period to compute the optimal strategy using the model. The optimal cardinality for this data is 11, and the optimal portfolio weights, along with the stock names of the optimal portfolio, are given in Table 12.3. This means the algorithm selects these 11 out of 50 stocks and their optimal allocation to achieve the maximum SR.

In this case, the SR is obtained as $SR = \dfrac{\rho - r_f}{w^T C w} = 0.1424$, considering the risk-free return $r_f = 0$. Information ratio is $\dfrac{\rho - \rho_I}{w^T \bar{C} w} = -0.1818$, where ρ_I corresponds to the BSE-SENSEX index. The beta for the optimal portfolio is obtained as 0.0516 by considering the market return for the BSE-SENSEX index.

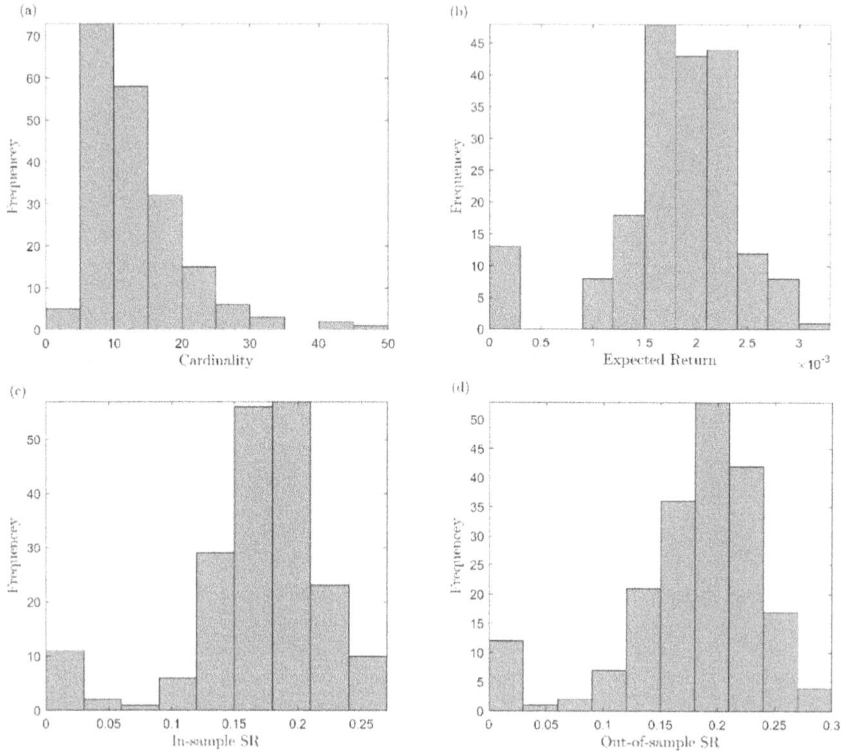

Figure 12.3 Distribution of the results in the sampling approach.

Table 12.2 Results of Various Measures in the Sampling Approach.

	Cardinality	*Expected Return*	*In-Sample SR*	*Out-of-Sample SR*
Mean	11	1.8×10^{-3}	0.1679	0.1808
Median	13.14	1.9×10^{-3}	0.1741	0.1939

Table 12.3 Optimal Portfolio Stock Weights

Stock Name	*Cyient*	*Punjab National Bank*	*Axis Bank*	*Marico*	*Godrej Consumer Products*	*Colgate-Palmolive*
Weight	0.0160	0.0002	0.2480	0.0520	0.0177	0.0401
Stock Name	Britannia Industries	Force Motors	Bajaj Auto	NTPC	Adani Power	
Weight	0.0269	0.0690	0.3460	0.1279	0.0560	

12.5 CONCLUSION

This chapter determines the optimal expected return and the optimal number of assets to be included in a portfolio to achieve maximum performance. The problem is modeled as a bilevel portfolio optimization model, where the upper level determines performance, while the lower level manages an efficient portfolio with cardinality constraints and expected return as parameters. It is proven that the stopping point of the method provides the optimal portfolio for the bilevel portfolio model. The method is tested on a set of stocks to validate the approach.

The computational efficiency of the proposed method is directly associated with the efficiency of the branch-and-bound technique. Like the branch-and-bound method for the integer programming, this algorithm works efficiently for the low-dimensional bilevel portfolio optimization models. By pruning unnecessary branches at different nodes, the algorithm can reduce the search space. These are the advantages of the algorithm. However, it has some limitations. Each node corresponds to the solution of a specific relaxed problem. If the number of assets is very large, these methods become computationally expensive, as the number of nodes may increase. In the worst case, this number is 2^n; hence, the number of nodes grows exponentially with the number of assets. The algorithm is valid as the lower-level problem of the proposed bilevel model has a unique solution. If the lower-level problem has a non-unique solution, then the construction of P_{R^i} will be challenging. Hence, in that case, the methodology and the problem P_{R^i} have to be modified to accommodate all the solutions of the lower problem, which is challenging, which may be considered as the extension of the present contributions.

Data availability: The data was collected from the site https://www.bseindia.com/markets/equity/EQReports/StockPrcHistori.aspx?flag=0&type=ETF.

REFERENCES

1. Morgan, J.: *Reuters. riskmetrics–technical document.* World Wide Web. http://www. riskmetrics. com/rmcovv. html (1996)
2. Lwin, K.T., Qu, R., MacCarthy, B.L.: Mean-var portfolio optimization: A nonparametric approach. *European Journal of Operational Research* 260(2), 751–766 (2017)
3. Rockafellar, R.T., Uryasev, S.: Conditional value-at-risk for general loss distributions. *Journal of Banking & Finance* 26(7), 1443–1471 (2002)
4. Konno, H., Koshizuka, T.: Mean-absolute deviation model. *IIE Transactions* 37(10), 893–900 (2005)
5. Konno, H., Wijayanayake, A.: Portfolio optimization problem under concave transaction costs and minimal transaction unit constraints. *Mathematical Programming* 89, 233–250 (2001)

6. Lobo, M.S., Fazel, M., Boyd, S.: Portfolio optimization with linear and fixed transaction costs. *Annals of Operations Research* 152, 341–365 (2007)
7. Yen, Y.-M., Yen, T.-J.: Solving norm constrained portfolio optimization via coordinate-wise descent algorithms. *Computational Statistics & Data Analysis* 76, 737–759 (2014)
8. Hooshmand, F., Anoushirvani, Z., MirHassani, S.: Model and efficient algorithm for the portfolio selection problem with real-world constraints under value-at-risk measure. *International Transactions in Operational Research* 30(5), 2665–2690 (2023)
9. Marhfor, A.: Portfolio performance measurement: Review of literature and avenues of future research. *American Journal of Industrial and Business Management* 6(4), 432–438 (2016)
10. Sharpe, W.F.: Capital asset prices: A theory of market equilibrium under conditions of risk. *The Journal of Finance* 19(3), 425–442 (1964)
11. Benita, F., L´opez-Ramos, F., Nasini, S.: A bi-level programming approach for global investment strategies with financial intermediation. *European Journal of Operational Research* 274(1), 375–390 (2019)
12. Leal, M., Ponce, D., Puerto, J.: Portfolio problems with two levels decision-makers: Optimal portfolio selection with pricing decisions on transaction costs. *European Journal of Operational Research* 284(2), 712–727 (2020)
13. González-Díaz, J., González-Rodríguez, B., Leal, M., Puerto, J.: Global optimization for bilevel portfolio design: Economic insights from the Dow Jones index. *Omega* 102, 102353 (2021)
14. Bayat, M., Hooshmand, F., MirHassani, S.A.: Optimizing risk budgets in portfolio selection problem: A bi-level model and an efficient gradient-based algorithm. *IISE Transactions* 56(8), 841–854 (2024)
15. Kobayashi, K., Takano, Y., Nakata, K.: Bilevel cutting-plane algorithm for cardinality-constrained mean-CVaR portfolio optimization. *Journal of Global Optimization* 81(2), 493–528 (2021)
16. Cesarone, F., Lampariello, L., Merolla, D., Ricci, J.M., Sagratella, S., Sasso, V.G.: A bilevel approach to ESG multi-portfolio selection. *Computational Management Science* 20(1), 24 (2023)
17. Chen, X., Kelley, C., Xu, F., Zhang, Z.: A smoothing direct search method for Monte Carlo-based bound constrained composite nonsmooth optimization. *SIAM Journal on Scientific Computing* 40(4), 2174–2199 (2018)
18. Jing, K., Xu, F., Li, X.: A bi-level programming framework for identifying optimal parameters in portfolio selection. *International Transactions in Operational Research* 29(1), 87–112 (2022)
19. Best, M.J., Grauer, R.R.: On the sensitivity of mean-variance-efficient portfolios to changes in asset means: some analytical and computational results. *The Review of Financial Studies* 4(2), 315–342 (1991)
20. Broadie, M.: Computing efficient frontiers using estimated parameters. *Annals of Operations Research* 45, 21–58 (1993)
21. Jorion, P.: Bayes-Stein estimation for portfolio analysis. *Journal of Financial and Quantitative Analysis* 21(3), 279–292 (1986)
22. Bauder, D., Bodnar, T., Parolya, N., Schmid, W.: Bayesian mean–variance analysis: optimal portfolio selection under parameter uncertainty. *Quantitative Finance* 21(2), 221–242 (2021)

23. Fernholz, E.R., Fernholz, E.R.: *Stochastic Portfolio Theory*. Springer, New York, NY (2002)
24. Moon, Y., Yao, T.: A robust mean absolute deviation model for portfolio optimization. *Computers & Operations Research* 38(9), 1251–1258 (2011)
25. Jorion, P.: Portfolio optimization in practice. *Financial Analysts Journal* 48(1), 68–74 (1992)
26. Fang, Y., Lai, K.K., Wang, S.: *Fuzzy Portfolio Optimization: Theory and Methods* vol. 609. Springer, Berlin, Heidelberg (2008)
27. Dempe, S.: *Foundations of Bi-Level Programming*. Kluwer Academic Publishers, Dordrecht (2002)
28. Dempe, S., Dutta, J.: Is bilevel programming a special case of a mathematical program with complementarity constraints? *Mathematical Programming* 131(1), 37–48 (2012) https://doi.org/10.1007/s10107-010-0342-1
29. Chang, T.-J., Meade, N., Beasley, J.E., Sharaiha, Y.M.: Heuristics for cardinality constrained portfolio optimisation. *Computers & Operations Research* 27(13), 1271–1302 (2000)
30. Fortuny-Amat, J., McCarl, B.: A representation and economic interpretation of a two-level programming problem. *Journal of the Operational Research Society* 32(9), 783–792 (1981)
31. Jiang, H., Ralph, D.: Smooth SQP methods for mathematical programs with nonlinear complementarity constraints. *SIAM Journal on Optimization* 10(3), 779–808 (2000) https://doi.org/10.1137/S1052623497332329
32. Kanzow, C., Schwartz, A.: A new regularization method for mathematical programs with complementarity constraints with strong convergence properties. *SIAM Journal on Optimization* 23(2), 770–798 (2013)

Chapter 13

On Some Perspectives of Gradient-Based Optimization Algorithms in Machine Learning

N. Munikumar, Balakrushna Tripathy, and R. Manjula

13.1 INTRODUCTION TO OPTIMIZATION

Optimization is a concept used widely in our day-to-day life, knowingly or unknowingly. We utilize several resources, which can be either natural or artificial. Since the resources are always finite, efficient use of the resources is very much required. This is where the role of optimization comes into play. The resources that are to be optimized are usually profit, quality, time, space, money, infrastructure available, etc. Optimization is of two kinds: maximization and minimization. There is a function involved in the process, given by an expression involving several variables. These variables are called decision variables. Since the resources are limited, several constraints arise and are associated with every optimization problem. A feasible solution to the problem is a set of values of the decision variables that satisfy the constraints. From among the feasible solutions, one must find out the ones that are also optimal; i.e., they make the value of the objective function maximum or a minimum.

The generalized structure of an optimization problem is

Maximize/Minimize the objective function, $f(\bar{x})$,
subject to certain constraints $g_i(\bar{x}) \le 0, i = 1,2,....m$.

It may be noted that in case there are some constraints that are of the form $h_j(\bar{x}) \ge 0$, these can be transformed to the $\le$ form. If there are any equality constraints, then those can be transformed into two constraints involving $\le$. The components of $\bar{x}$ are the decision parameters. The values $\bar{x}$ that satisfy all the constraints are called feasible solutions and form the solution space for the problem at hand. We need to find the optimal solution(s). There are many methods that are applicable to only specific cases, like the graphical solutions.

The exponential growth in the amount of data has attracted several researchers to use optimization in the fields of machine learning (ML), as the increase in data size leads to an increase in the complexity of the models. There are several challenges faced by the researchers in this field. However,

DOI: 10.1201/9781003642091-13

due to their continuous efforts, many methods for solving optimization problems and generating their solutions have been found.

13.2 MACHINE LEARNING BASICS

13.2.1 What Is Machine Learning?

ML is a subfield of artificial intelligence (AI), in which the agent is told what is to be found but not the process to obtain it. The agent learns it from the dataset through several interactions and finds patterns and rules associating the elements. Statistics, when compounded with the field of computer science, generate ML. It can be said that the basic statistics are the core of ML, and the corresponding algorithms are obtained from statistical equations framed thereof. The approach mentioned above is called the data-driven approach, and ML comprises specific techniques that allow machines to learn and improve their performance. An ML model can be considered as an algorithm that finds patterns by interacting with the given data such that the knowledge obtained helps in making decisions and predictions. This turns out to be an automatic process and forms the aim of ML. After proper framing, the model can predict unknown data without any fault. From the beginning, human beings in their daily lives are helped by the efficiency of machines, and so ML as a theory drives it further in a systematic manner. To achieve this goal, machines are required to think like humans and have a clear understanding to mimic human beings in making decisions.

13.2.2 Role of Optimization in Machine Learning

Optimization algorithms are crucial in the field of ML as they help in improving model efficiency and accuracy by minimizing the loss or error function through solving the objective function. Optimization is a core component of the model training process for reducing the errors or deviations in the training data and generalizing the model more effectively to unseen data. These algorithms are not only helpful for boosting performance by reducing errors, but these are also essential for handling several other issues like overfitting, scalability, and computational efficiency. Optimization and ML exist together and are mutually reinforcing, with progress in one driving advancements in the other. As mentioned, optimization is the core and essential component of ML for optimizing (minimizing) the loss/error functions impacting the efficiency and accuracy of ML models [1]. With the recent advancements in ML techniques, optimization algorithms are used in data processing and extracting valuable insights from large datasets, making the optimization extremely important with growing data complexity and volume. In practice, various optimization algorithms are integrated and made available with various ML frameworks and libraries, allowing their implementation in real-world applications. The selection of the right optimization algorithm for ML can greatly influences their performance and efficiency [2].

13.2.3 Types of Learning

Learning is the process of knowledge acquisition, skill development, or enhancing understanding. It can be divided into two categories: learning from our day-to-day life, which is added to our experience, or learning from systematic courses. In the ML domain, the learning processes are categorized into several types. We discuss below some of the popular approaches to learning.

13.2.3.1 Supervised Learning

Supervised learning techniques require a good number of labeled training samples/data provided during training to make predictions. It involves learning from a dataset where both the inputs and their corresponding outputs are known, which allows the model to learn and find a relationship between them. The model can predict outputs for fresh data that are not seen earlier, as it has learned patterns and relationships among the input data during the training phase, as illustrated in Figure 13.1. Once a mapping is generated that gets the labeled input data and returns the output, the purpose is achieved. Usually, the input vector represents the training data, and the labels or tags form the output data.

An output label from vector *Y* is the explanation of its respective input example from input vector *X*. The input vector *X* and the output vector *Y* together form a training example. In other words, training data comprises training examples. All the samples in the training data are associated with corresponding label(s) and form the output vector *Y*. Sometimes human beings label the datasets as supervisors. But, conveniently, machines are used for this work. This makes the work efficient and unbiased.

During the training process, the input vector generates an output that is compared with the desired output vector. If the actual output and desired output do not match, an error signal is generated by the network. This error signal is used for the adjustment of parameters until the actual output matches the desired output.

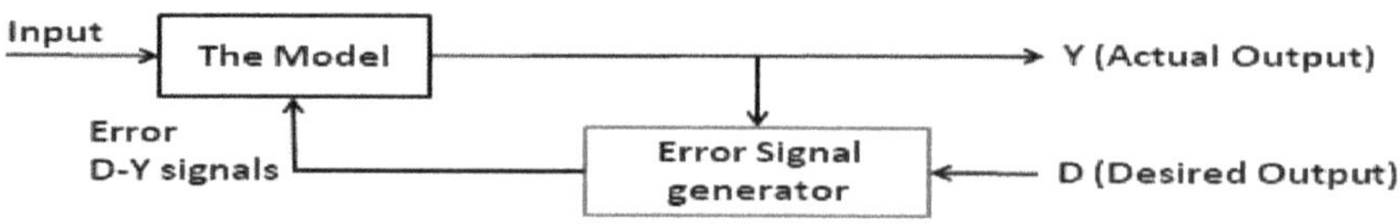

Figure 13.1 Supervised learning training process.

13.2.3.2 Unsupervised Learning

Data that do not have any labels are used for unsupervised learning. The training is done by observing the patterns and regularities in the input data. As expected, there are no supervisors, and the input data is taken without any labels and processed as shown in Figure 13.2. Basically, data elements are grouped based on their similarity. The elements that are more like each other form a single group. The elements in different groups are less like each other. The groups are called clusters. Unlike supervised learning, there is no output vector Y as the data are unlabeled. There are several measures of similarity. The Euclidean distance is the most popular similarity metric.

There are various ML algorithms that take care of the clustering process. The earliest and most popular one is the classical K-means algorithm. Here, the algorithm requires the number of clusters to be formed by the user and forms those many clusters at the end. There is no human intervention except for providing the number of clusters required. Sometimes the clusters to be formed are not disjoint from each other, and some common elements are involved with their belonging to the clusters to some degree. These are based upon the uncertainty-based models, and several such models and algorithms are in existence in the literature. Unlabeled data does not have any predefined categories, and as such, the algorithm must take care of finding patterns inside the input data and finding relationships among them. It is a work that is challenging and at the same time interesting. Rather, this process provides interesting properties of the data elements and insights.

During the training process, the model identifies the input patterns and organizes these patterns into clusters. In case no class is found with any pattern, a new class is formulated. For every input data, the model generates a class that identifies the pattern to which it belongs.

13.2.3.3 Semi-Supervised Learning

As the name suggests semi-supervised learning is a combination of labeled and unlabeled data, with a large portion of them being unlabeled and the rest being labeled. As usual, the aim of this type of learning is the prediction of new and unseen data accurately. In the case when it is expensive or takes too much time to label the data, this learning method is very effective. It can be considered as a method that narrows the gap between supervised

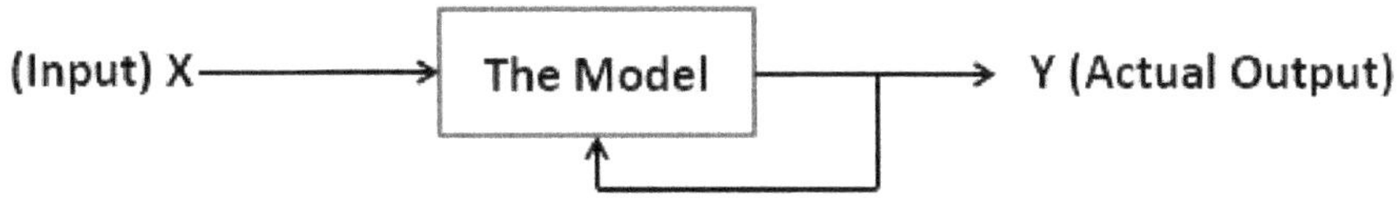

Figure 13.2 Unsupervised learning training process.

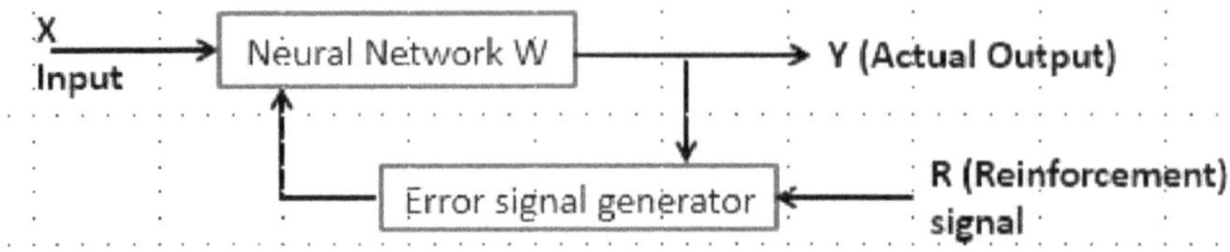

Figure 13.3 Reinforcement learning process.

learning and unsupervised learning, and when there is a large amount of unlabeled data in the input.

13.2.3.4 Reinforcement Learning

The latest introduction into the realm of ML is reinforcement learning (RL). The learning process under RL is a continuous process, and it gets better. An agent takes care of the process by interacting with the environment, getting feedback in the form of rewards and penalties. Beginning with the start state, at every step, there are several actions available to the agent. In a specific state, every action has specific feedback, as shown in Figure 13.3. The goal is to gather reward points as much as possible and maximize the cumulative reward points. Here, the outputs are not labeled but are graded. The agent discovers a policy through interaction with the environment, obtaining knowledge about different actions through trial-and-error methods, receiving rewards or penalties.

13.2.4 Machine Learning Use Cases

13.2.4.1 Regression

Regression is one of the statistical techniques that primarily aims to predict continuous outputs by modeling the relationship between one dependent variable and one or more independent variables. This technique is widely used across several industries, such as finance, healthcare, and engineering, to predict/forecast trends and support decision-making. In ML, regression techniques fall into the category of supervised learning, which trains the model to estimate the dependent variable for a set of independent variables. This technique is heavily used in applications such as stock price prediction, selling price estimation, and weather forecasting. The selection of an appropriate regression algorithm determines the accuracy and computational efficiency of predictions. For evaluating the performance of regression algorithms, metrics like mean squared error (MSE), mean absolute error (MAE), and *R*-squared error [3] are used. Linear regression is a popular algorithm that comes under regression and is categorized under supervised learning. In this process, the most fit linear function is generated from the datasets,

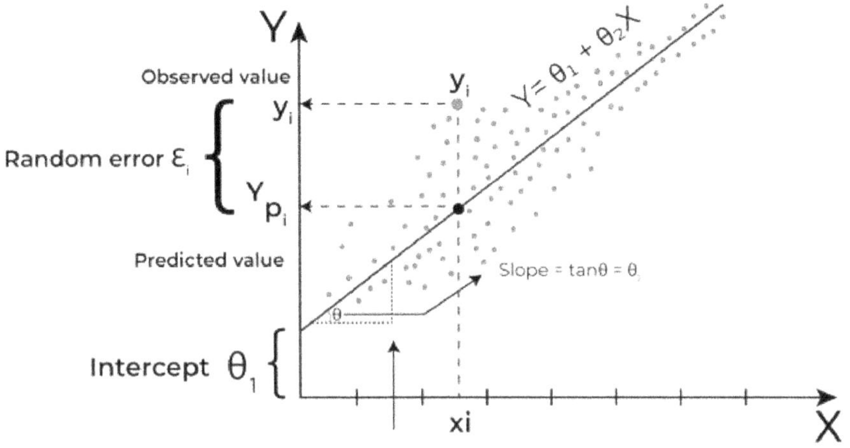

Figure 13.4 Best line, *X*-axis, the work experience, and *Y*-axis, the salary of a person.

and this helps in predicting new and unseen data elements, as shown in Figure 13.4. Perhaps the simplest version of linear regression occurs when there is only one independent feature, termed as simple linear regression, and is given by the equation

$$y = \alpha_1 x + \alpha_0 .$$

The linear regression objective is to find a straight line that best fits the data provided in the best possible way, in the sense that the deviations between the actual value and the predicted value are at a minimum. In turn, the line generated defines how the dependent and independent variables are related to each other. The gradient of this line measures the rate of change in the value of the dependent variable for a change in the independent variable by a unit.

13.2.4.2 Classification

The classification technique also falls into the category of supervised learning in ML, and it enables the models to learn or identify patterns from existing labeled examples and generate accurate predictions on unseen data. Like any ML algorithm, the effectiveness of a classification model lies with the quality of the data supplied as part of training. In classification model performance, the essential preprocessing steps, like managing missing values and applying feature scaling, play a crucial role in enhancing the model performance. The capability of pattern identification in classification models is exhibited across various domains, like email spam detection, fraud detection, medical diagnosis, and churn prediction for informed decision-making. Some of the popular classification algorithms most widely used in a variety of domains include logistic regression, decision trees, random forest, and support vector machine (SVM). Classification tasks may be of different types:

- Binary Classification: The input data is classified into two categories (e.g., spam/ham or yes/no).
- Multi-class Classification: Here, the input data is classified into multiple classes (e.g., identifying different types of objects).
- Multi-label Classification: In this classification approach, a collection (multiple) of labels is assigned to a single data point.
- Imbalanced Classification: The classification approach in which the datasets contain unequal or imbalanced training data for each class label, i.e., more positive samples and fewer negative samples, or vice versa (e.g., fraud detection dataset where the fraud cases are fewer in number compared to the benign).

13.2.4.3 Clustering

Clustering is another important technique in ML that falls under the unsupervised learning category. As it is an unsupervised technique, no training data is supplied for this algorithm. This technique clusters/groups similar data points based on their similarities in the attribute values. These algorithms help uncover the underlying patterns in the data, particularly helpful in the case of high-dimensional spaces, to identify relationships among samples. The popular clustering algorithms widely used in real-world applications include *K*-means, hierarchical clustering, mean-shift, and density-based spatial clustering of applications with noise. With this wide range of real-world applications, clustering acts as a powerful tool in statistical data analysis.

13.2.4.4 Applications for Clustering

- **Financial Markets:** One of the major applications of clustering techniques is in the financial markets, toward applications such as time series analysis, portfolio optimization, and risk management. These techniques uncover the hidden patterns in financial data, enhance the decision process, and extract valuable insights.
- **Social and Political Research:** With the social media data, these techniques uncover the hidden patterns and offer valuable insights into social behaviors and political trends.
- **Healthcare:** In healthcare, the clustering technique analyzes patient data and predicts diseases by using healthcare resources efficiently. As healthcare is high-dimensional data generated from a variety of sources, clustering is very helpful for working with electronic health records.

13.2.5 Classical Adaptive and Momentum-Based Methods

For any ML-based applications, gradient-based optimization methods play a pivotal role in providing efficient solutions through minimizing the objective function or loss function and improving the model performance. Over the past two decades, optimization research has developed various gradient descent optimization algorithms for ML tasks. The popular approaches include classical gradient descent (GD), batch gradient descent (BGD), and stochastic gradient descent (SGD). These variants suffer from various critical challenges, like slow convergence and sensitive hyperparameters in ML applications. Furthermore, the recent advancements in this domain include adaptive and momentum-based GD algorithms, which have enhanced the convergence speed and stability in the case of real-world high-dimensional data spaces.

Adaptive methods, including AdaGrad, RMSProp, and Adam, are developed to dynamically adjust learning rates with the aim of improving convergence in non-convex optimization problems. Adam is known to be very popular for its balance of time efficiency and accuracy. Momentum-based methods enhance the rate of convergence by incorporating momentum that helps in navigating high-dimensional spaces more effectively. The most popular momentum-based method is the Nesterov accelerated gradient (NAG), which enhances convergence speed by incorporating momentum that helps in navigating higher-dimensional spaces more effectively. Another momentum-based method is the variation of Adam, i.e., Nadam, which integrates momentum with Adam to exhibit superior performance in terms of accuracy across various datasets. In addition to the above classical adaptive and momentum-based methods, we discuss the most advanced optimization methods that combine extended research based on gradient-based optimization algorithms, which consider exponential smoothing, gradient accumulation, energy-based strategies, along with momentum, by highlighting their importance, application, and advancements.

Razzouki et al. [4] provided a comprehensive overview of advanced gradient methods used in ML for optimizing model training. Furthermore, they discussed the importance of gradient-based methods as these are essential for minimizing loss functions and enhancing model performance across various ML tasks. The authors extended the discussions from classical methods, adaptive methods, to momentum-based gradient methods. In summary, this article provided a detailed survey of various gradient optimization techniques, outlining their strengths, limitations, and the ongoing evolution in the field to tackle complex optimization challenges in ML. Maurya and N. Yadav [5] compared the widely used first-order gradient-based optimization algorithms that include SGD (SGD with momentum), AdaGrad (adaptive gradient), AdaDelta (adaptive delta), RMSProp (root mean square propagation), Adam (adaptive moment estimation), Nadam (Nesterov accelerated adaptive moment estimation), and AdaMax (maximum adaptive moment estimation). For the evaluation of these optimization strategies, popular ML techniques such as regression, binary classification, and multi-classification are used with randomly selected databases. It is observed that Nadam outperformed other optimization methods in terms of correctness across datasets, and the Adam optimizer exhibited the best performance regarding time efficiency and accuracy. Overall, this comparative study discussed the strengths and weaknesses of each optimization algorithm.

Sakovich et al. [6] presented an advanced gradient optimization method called momentum-based adaptive momentum gradient descent optimization. This method incorporates exponential decay and an adaptive learning rate based on the discrete second-order derivative of gradients. This method is primarily used for neural network training. This optimizer was evaluated on various optimization problems, including the minimization of multivariate real functions and function approximation using multilayer neural networks. The results indicated effective accumulation of gradient accumulators and showcased the optimization performance. Tian [7], in his article, addressed two significant challenges in ML and data science, namely, efficient optimization methods for large-scale models and the discovery of dynamical systems from observational data. By utilizing an adaptive-energy-based strategy with stability irrespective of learning rate, the authors proposed a new set of gradient-based optimization algorithms. The proposed new algorithm is referred to as stochastic gradient with energy and momentum (SGEM), which incorporates momentum acceleration and demonstrates rapid convergence, superior generalization capabilities in various experimental results on deep learning applications. In summary, this article contributed to theoretical understanding and practical applications in large-scale optimization problems. Rao et al. [8] presented a gradient-based optimization method that primarily aimed to suppress interference in linear arrays through the control of electrical parameters, specifically focusing on amplitude-only and phase-only adjustments. The authors employed a gradient-based optimization algorithm (GOA), which is recognized for its efficiency in solving optimization problems typically addressed by evolutionary algorithms. The primary objective of the optimization method is to maximize the main beam gain while simultaneously minimizing the peak sidelobe level by adhering to null constraints.

Gu and Chen [9] discussed a novel gradient-based approach for multilevel optimization and addressed various challenges in efficiently managing nested structures in ML applications, such as hyperparameter tuning and continual learning. The authors presented an automatic differentiation method specifically designed for gradient computation in trilevel optimization problems. This chapter also emphasized the algorithm capabilities to provide a general version of implicit differentiation and contributed significantly to the field of multilevel optimization by offering a more efficient and accurate method for handling complex optimization problems. Andel Finger [10] discussed the application of differentiable agent-based simulation in the context of gradient-guided simulation-based optimization. Various traditional simulation-based optimization methods assumed that the gradient of simulation with respect to input cannot be directly evaluated, which necessitated for gradient estimation methods, especially in high-dimensional input spaces. This research mainly focused on time-driven agent-based solutions, in which common discrete model elements are replaced with smooth approximations. This study also examined specific applications like microscopic traffic models and epidemic models. The results indicated that gradient-based optimization methods significantly outperformed gradient-free methods in traffic signal timing optimization problems of high input dimensions.

13.3 GRADIENT-BASED OPTIMIZATION METHODS

For any ML or data-driven tasks, gradient-based optimization methods are crucial in enabling efficient training of models by adjusting the parameters iteratively to minimize a loss function. The methods use the gradient to find local minima or maxima for any function, and they primarily rely on the calculation of gradients for finding optimal solutions. Furthermore, these methods are used in handling large-scale optimization problems by adapting to large real-time scenarios. Although these optimization methods are highly effective, they do suffer from certain limitations, like getting trapped in local minima, sensitivity to hyperparameter settings, and computational complexity in high-dimensional spaces. The choice of an appropriate method depends on the problem context, establishing a balance among speed, accuracy, and stability. Continuous advancements and research in this field are improving these techniques to fit the needs of ML professionals.

The "order" of an optimization plays a crucial role in the optimization (either maximize/minimize) of the objective function. The order indicates the amount of information about the objective function the algorithm uses to guide its search. The classification of algorithms on the basis of their convergence behavior and computational complexity depends upon their order. Fundamentally, the algorithms can be classified into three categories as far as their orders are considered. These are as follows:

- Zeroth order
- First order
- Second order.

Zeroth-order gradient optimization, also called the derivative-free or black-box optimization, depends only upon function values and not gradients. These are useful when gradients are unavailable or expensive to compute. These methods can handle noisy, non-differentiable, or discontinuous functions but are slower in convergence and less efficient for smooth problems. Some such methods are as follows:

- Random search
- Genetic algorithms
- Simulated annealing
- Bayesian optimization.

In this work, we are not focused on discussing these algorithms. Rather, we shall focus on the other two types of order algorithms, namely, first-order optimization algorithms and second-order optimization algorithms.

13.3.1 First-Order Optimization Algorithms

First-order optimization algorithms are, as the name indicates, optimization algorithms that use the first-order derivative of the objective function to find the minimum or maximum value. These algorithms solely rely on the gradient (derivative) to update the learning parameters iteratively. This makes them more efficient for high-dimensional optimization problems. The key characteristics of the first-order optimization methods are as follows:

- Only gradient (first derivative) information is used;
- Suitable for large-scale optimization problems;
- Computationally less expensive than second-order or higher-order optimization methods;
- Do not require second-order derivatives (Hessian matrix).

Based on how the parameters get updated during optimization, the first-order optimization algorithms are classified into three types:

- GD algorithms
- Momentum-based algorithms
- Adaptive learning rate algorithms

Since first-order optimization methods use gradient information only, the computational cost is lower, the memory usage is less, and the convergence speed is slower and scalable. The major applications of first-order optimization methods include training deep learning models, large-scale ML applications, and optimization in higher-dimensional spaces.

13.3.1.1 Gradient Descent Algorithms

In general, a gradient (derivative) is a linear approximation of a function. Local maximum or local minimum of an objective function is determined by the GD iteratively at its points of convergence. The neural networks and other ML models are usually trained through the GD algorithm [11]. In this algorithm, the magnitude and the direction are reflected in the gradient. This algorithm aims to move toward the set of optimal values of the parameters, and the movement is in the opposite direction of the gradient. It also fits the training data in the best way.

At each iteration, the cost function is stepped down in the same direction as that of the steepest descent, as shown in Figure 13.5. The values of the parameters to match with this direction, the minimum values of the cost function, are sought by the gradient descent, and as such, the values of the parameters generated provide the optimal values. The steps should be appropriately maintained such that there is no skipping of the minimal points. The parameter α controls these steps, and it is also called the learning rate. The values θ_i increase if the gradient value is negative and decrease if the gradient value is positive.

Selection of Learning Rate:

For the convergence of the gradient in a reasonable time, the selection of a suitable learning rate is very important. The large values α may lead to long steps, and furthermore, the appropriate value of the minimum point may be overstepped. On the other hand, very small values α may lead to insignificant steps, and the process of convergence will take a much longer time.

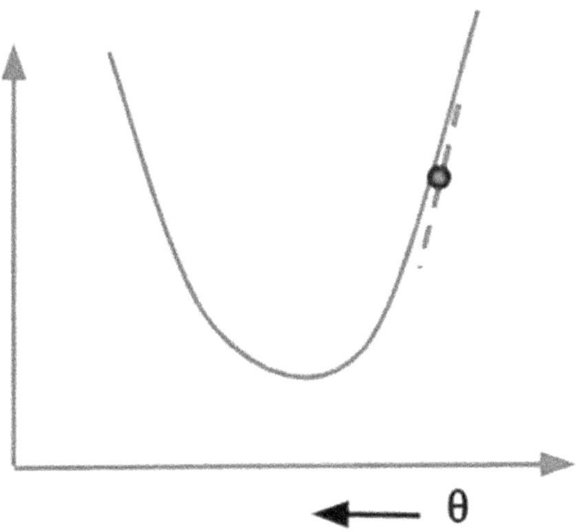

Figure 13.5 Relationship between gradient and θ.

Derivation:

Let the cost function J be defined in terms of two variables θ_1 and θ_2, i.e., $J(\theta_1, \theta_2)$. Now, differentiating the cost function J with respect to θ_1, we get

$$\frac{\partial J(\theta_1,\theta_2)}{\partial \theta_1} = \frac{\partial}{\partial \theta_1}\left[\frac{1}{n}\left(\sum_{i=1}^{n}\left(\vec{y_i} - y_i\right)^2\right)\right] = \frac{1}{n}\sum_{i=1}^{n} 2\left(\vec{y_i} - y_i\right)^2 \frac{\partial}{\partial \theta_1}\left(\vec{y_i} - y_i\right)$$

$$= \frac{1}{n}\sum_{i=1}^{n} 2\left(\vec{y_i} - y_i\right)\frac{\partial}{\partial \theta_1}\left(\theta_1 + \theta_2 x_i - y_i\right) = \frac{1}{n}\sum_{i=1}^{n} 2\left(\vec{y_i} - y_i\right) = \frac{2}{n}\sum_{i=1}^{n}\left(\vec{y_i} - y_i\right)$$

Now, differentiating the cost function J, with respect to θ_2, we get

$$\frac{\partial J(\theta_1,\theta_2)}{\partial \theta_2} = \frac{1}{n}\sum_{i=1}^{n} 2\left(\vec{y_i} - y_i\right)\frac{\partial}{\partial \theta_2}\left(\theta_1 + \theta_2 x_i - y_i\right) = \frac{1}{n}\sum_{i=1}^{n} 2\left(\vec{y_i} - y_i\right)x_i$$

$$= \frac{2}{n}\sum_{i=1}^{n}\left(\vec{y_i} - y_i\right)x_i$$

Using this, the coefficients of the best-fitting line to represent the training data associated with the objective function can be deduced. The gradient is moved in the direction of the MSE and changes occur in the coefficients, and depending on these, the learning rate α and the corresponding intercept can be obtained as shown in Figure 13.6.

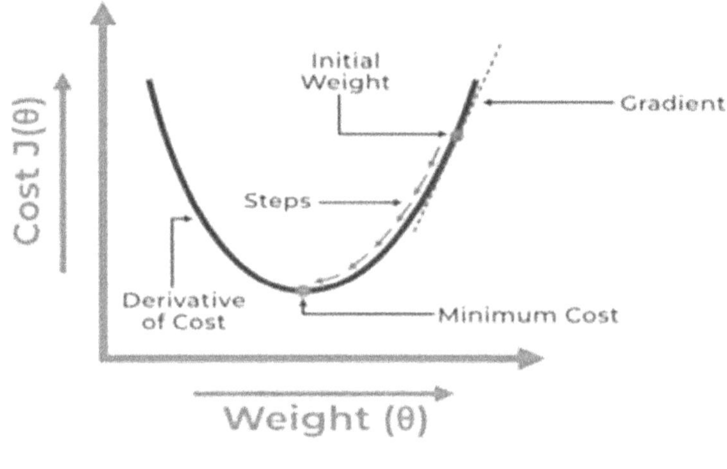

Figure 13.6 Relationship between θ and $J(\theta)$.

$$\theta_1 = \theta_1 - \alpha \frac{\partial J(\theta_1, \theta_2)}{\partial \theta_1} = \theta_1 - \alpha \left(\frac{2}{n} \sum_{i=1}^{n} \left(\overrightarrow{y_i} - y_i \right) \right)$$

$$\theta_2 = \theta_2 - \alpha \frac{\partial J(\theta_1, \theta_2)}{\partial \theta_2} = \theta_2 - \alpha \left(\frac{2}{n} \sum_{i=1}^{n} \left(\overrightarrow{y_i} - y_i \right) x_i \right)$$

The update rule is formulated as

$$\theta_{t+1} = \theta_t - \eta \nabla f(\theta_t). \tag{13.1}$$

$\theta_t \rightarrow$ Current parameter values at iteration t
$\eta \rightarrow$ Learning rate (step size)
$\nabla f(\theta_t) \rightarrow$ Gradient of the loss function $f(\theta)$ with respect to θ_t
$\theta_{t+1} \rightarrow$ New/updated parameter value
Generally, the GD algorithm involves the following steps.

- **Step 1:** Initialize parameters:
  - To begin with, initialize the parameters with random values, use these values to estimate the model, and obtain the value of the cost function.
- **Step 2:** Compute the gradient
  - The gradient $\nabla f(\theta_t)$ is the first derivative of the loss function with respect to the parameters.
  - This represents the direction of the steepest increase in the function.
- **Step 3:** Move in the opposite direction
  - As the objective is to minimize the function, we must subtract the gradient from the current parameter values.
- **Step 4:** Control step size with learning rate η
  - The learning rate η controls how large the update step is.
  - For larger values of η, the updates may overshoot the minimum.
  - For lower values of η, convergence takes time, making the algorithm slow.
- **Step 5:** Iterate the process
  - This process continues until the change in parameter values is very small, which indicates that we are near the convergence (i.e., near to local or global minimum).

The GD algorithms incur huge computational costs for processing the above algorithm. For example, let's calculate the computation cost over each iteration of the algorithm. Assume that we have 10,000 data points and 10 features for each data point. In this case, the sum of squared residuals will be the same as that of the data points, i.e., 10,000. To compute the gradient of the objective function with respect to each of the features, we need 10,000 * 10 = 100,000 computations per iteration.

In general, we use 100–1,000 iterations to train the model. In the most feasible case of 1,000 iterations, we have 100,000 * 100 = 100,000,000 computations to complete the algorithm. This calculation of GD will be an overhead; hence, GD is slow on huge datasets. So, the SGD (random gradient descent) is developed to overcome this high computational cost for large datasets. In stochastic gradient, the core idea is that we introduce randomness at each step of derivative calculation, and this method picks one data point randomly from the whole dataset to reduce enormous computational complexity. The other variant of the GD algorithm selects a small number of data points instead of just one point at each step, and this is called "mini-batch GD." This mini-batch GD strikes a balance between the goodness of the GD and the speed of SGD. These variants of GD algorithms will be discussed in detail below. The GD has three variants [12] based on the amount of data used to compute the gradient of the objective function.

13.3.1.2 Batch Gradient Descent

The batch GD computes gradients of the entire dataset for parameter updating per iteration. The gradients for all the data points in the training set are averaged, and the parameters are updated accordingly. The parameter updation happens only after the evaluation of all the training samples in the iteration.

The parameter update rule remains the same:

$$\theta_{t+1} = \theta_t - \eta \nabla f(\theta_t).$$

The advantage of this method is that it generates a more accurate value for the gradient, i.e., stable parameter updates and accurate convergence. The disadvantage on the other side is that this approach is computationally too expensive for larger datasets.

13.3.1.3 Stochastic Gradient Descent

One of the most popular algorithms, called SGD, is used in various ML applications and forms the basis of many neural networks. In general terms, a gradient is known as the slope of a surface. So, GD literally means descending a slope to reach the lowest point on that surface. For example, consider a parabolic curve $y = x^2 - 2x - 3$, shown in Figure 13.7.

When $x = 0$, the value of $y = -3$, and for $y = 0$, the equation $x^2 - 2x - 3 = 0$ may be expressed as the product of $(x-3)(x+1)$, resulting in the roots $(x = 3)$ and $x = -1$. From the above graph, we observe that the lowest point on the parabola occurs at $x = 1$. Here, the GD algorithm's objective is to find the value of "x" such that "y" is minimum, where y is the objective function that GD acts upon, to descend to the lowest point. The GD algorithm is expressed as an iterative algorithm that starts from

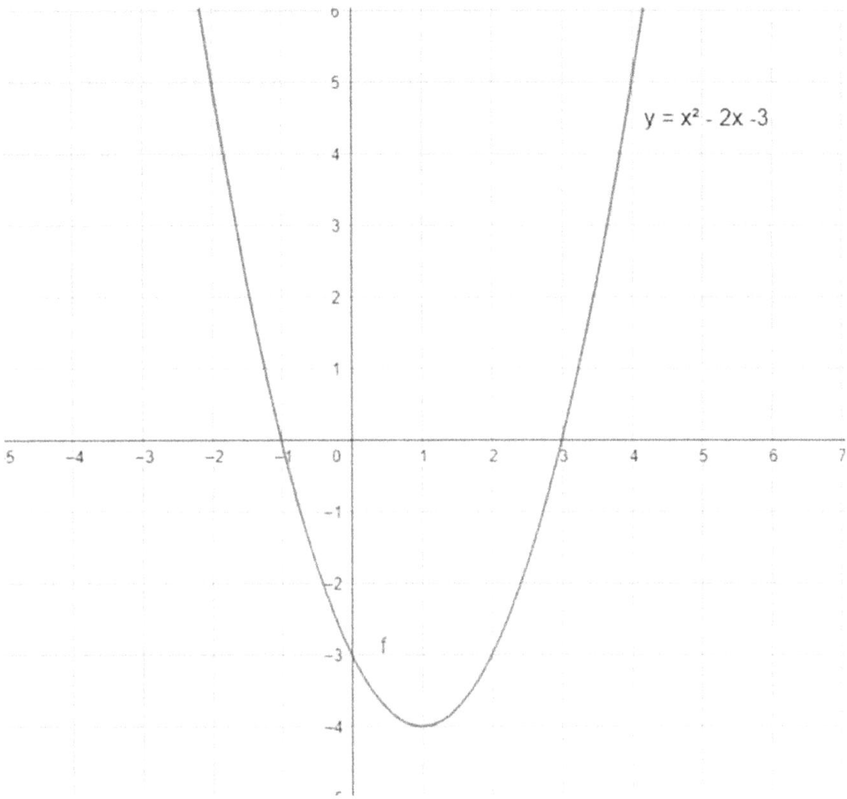

Figure 13.7 Graph of the function $y = x^3 - 2x - 3$.

a random point on a function and travels down its slope in steps until it reaches the lowest point of that function. This algorithm is useful in cases where the optimal points cannot be found by equating the slope of the function to 0. The general idea here is to start with a random point and find a way to update this point with each iteration such that we descend the slope.

The SGD algorithm is expressed in steps as follows:

Step 1: Calculate the gradient (slope) of the objective function with respect to each feature.
Step 2: Generate a set of random values as initial values for the parameters.
Step 3: Differentiate the objective function "y" with respect to "x."
Step 4: In case of more features like x_1, x_2, x_3, etc., calculate the partial derivative of "y"
with respect to each of the features.
Step 5: By substituting the parameter values, update the gradient function.
Step 6: Update the step sizes for each feature as

step size = gradient * learning rate.
Step 7: Update the new parameter values as
new parameters = old parameters – step size.
Step 8: Repeat Steps 3–7 until the gradient is almost 0.

SGD is a minified version of GD, addressing the main challenge of reducing the computational cost. In SGD, the gradient is computed for one of the randomly selected partitions in the shuffled dataset during each iteration, instead of the entire dataset, which significantly reduces computational time. The parameter update rule is slightly modified as

$$\theta_{t+1} = \theta_t - \eta \nabla f_i(\theta_t),$$

where $f_i(\theta_t)$ is the loss computed for the single selected sample. The advantage of this approach is faster updates, reduced computational costs, and suitability for large datasets. The disadvantages remain as noisy updates, as only one sample is considered and may not converge smoothly.

13.3.1.3.1 Mini-Batch Gradient Descent

The mini-batch GD is introduced to balance between the batch gradient and SGD methods. Here, the data points are split into small batches, and the gradients are computed. The parameters are updated after each batch in the iteration. This approach balances efficiency with reduced computational cost compared to batch GD and better convergence compared to SGD. But here, the batch size will be a hyperparameter that needs to be tuned as per the required metrics.

The parameter update rule is modified as

$$\theta_{t+1} = \theta_t - \eta \frac{1}{m} \sum_{i=1}^{m} \nabla f_i(\theta_t). \tag{13.2}$$

The advantage of this approach is that it balances both efficiency and stability but requires additional tuning of the batch size as a hyperparameter for quicker convergence.

As mentioned earlier, GD is one of the popular optimization techniques that is currently being used in various ML algorithms, including linear regression, logistic regression, various neural networks, and SVMs. Although GD is widely used in the training of ML models, it still possesses open challenges [12] that need to be addressed. The major challenges related to learning rates include the following:

- Determining the right learning rate is often tricky.
- A uniform learning rate is applied across all parameter updates.

13.3.1.4 Momentum-Based Methods

In gradient-based optimization methods, the term momentum [13] is a technique that accelerates the SGD toward the convergence of optimization tasks for the given objective function. This method updates the parameters by combining the current gradient with a fraction of the past gradient. Momentum uses a velocity vector, which is a moving average of the gradients. Momentum optimization updates weights by considering both the current gradient and a portion of the previous gradients. This approach reduces oscillations near the optimum, promoting smoother and quicker convergence. However, momentum can still struggle when dealing with complex error landscapes that contain numerous valleys and peaks. The major benefits of momentum include faster convergence, reduced oscillations, and improved stability compared to the gradient-based optimization methods.

13.3.1.4.1 Momentum Stochastic Gradient Descent

Momentum stochastic gradient descent (MSGD) [14] is a variant of a momentum-based optimization method with SGD that enhances convergence by adding a momentum coefficient, i.e., a fraction of the previous update to the current one. This generates a smooth optimization trajectory and reduces oscillations, as shown in Figure 13.8. The main intent of using momentum in gradient-based optimization is to accelerate the convergence and improve stability for high-dimensional datasets. This is achieved by accumulating the velocity vector in directions of consistent GD, which reduces oscillations and prevents the system from getting stuck in a local minimum. Momentum has proved to be very crucial in today's optimization techniques. It provided significant improvements in training the ML models by addressing a few of the limitations in both batch GD and SGD methods.

The update rules are

$$v_t = \beta v_{t-1} + (1-\beta) g_t, \theta_{t+1} = \theta_t - \eta v_t, \tag{13.3}$$

where η is the learning rate,

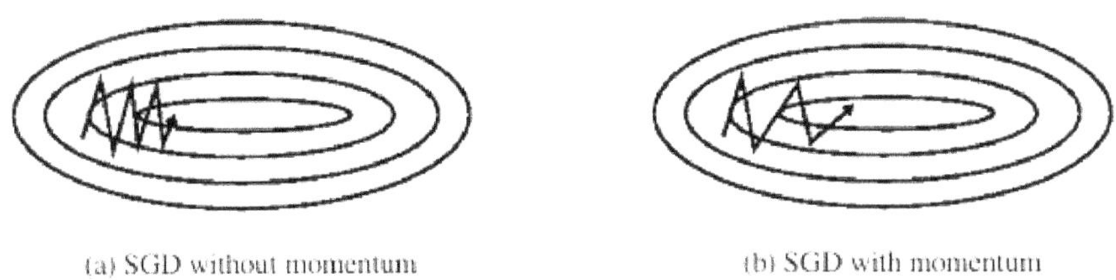

Figure 13.8 Momentum and learning rate adaptation.

v_t is the velocity at time step t,
β is the momentum term (typically between 0.9 and 0.99),
g_t is the gradient at time step t.

By using the momentum coefficient, the gradients of those dimensions that point in the same direction are increased, and updates are reduced. This results in increased convergence and reduced oscillations.

13.3.1.4.2 Nesterov Accelerated Gradient

Nesterov accelerated gradient (NAG) [14] is an improved variant of MSGD, aimed at achieving smoother and faster convergence. It anticipates the future gradient by looking ahead, allowing for more accurate adjustments that reduce overshooting and enhance stability. The update rules are

$$v_t = \beta v_{t-1} + \eta \nabla_\theta J(\theta_t - \beta v_{t-1}), \theta_{t+1} = \theta_t - v_t, \quad (13.4)$$

where η is the learning rate,
v_t is the velocity at time step t,
β is the momentum term (usually between 0.9 and 0.99),
$\nabla_\theta J(\theta_t - \beta v_{t-1})$ is the gradient at the "look ahead" position.

In other words, NAG computes the gradient at a "look-ahead" point instead of the current position, and it helps find the best solution faster and more accurately by looking ahead at future changes by reducing the chance of going too far. Hence, this optimization method works well for complex problems like deep learning [15].

13.3.1.5 Adaptive Learning Rate Methods

Adaptive learning rate methods are the optimization techniques that automatically adjust the learning rates during the optimization process based on the training progress. This technique is currently being used in ML, where learning rates are adjusted based on the progress of the weight updates. Furthermore, these methods also reduce the need for additional hyperparameter tuning as the learning rate gets updated automatically. The adaptive learning rate helps in improved convergence, improved generalization, reduced hyperparameter tuning, and stability compared to fixed learning rate methods.

13.3.1.5.1 AdaGrad (Adaptive Gradient Algorithm)

AdaGrad [16] is the most popular optimization algorithm with the capability to automatically adjust the learning rate for each parameter based on the sum of previous squared gradients. Learning rates are adaptive because the actual rate is determined from parameters. The parameters with a high gradient will have a reduced learning rate, and the ones with smaller gradients will get an increased learning rate. This feature of adjusting learning rates for each parameter allows this method to adapt to various tasks with sparse data possessing variations/changes in the parameters. AdaGrad is most beneficial to applications that require different learning rates for different parameters, eliminating the need for manual learning rate scheduling. The major disadvantage of this method is the aggressive decay of the learning rate, which results in prematurely stopping learning during long-term training cases.

13.3.1.5.2 Working Principle of AdaGrad

The AdaGrad method works in two steps:

Step 1: Gradient Descent: The GD method works by updating each of the weights, w, based on the gradient of the loss function, which indicates the direction to minimize error. During each iteration, the weight value "w" is updated by subtracting the product of the learning rate with the gradient, i.e., learning rate x gradient.

Step 2: Adaptive Learning Rate: AdaGrad modifies adaptive learning rates for each weight. To accumulate the gradient of each parameter, this method divides the learning rate by the square root of the sum of the past gradients squared. This modification reduces the parameters' learning rates having high gradients, allowing for more focus on features.

$$\text{The update rule is: } w_{t+1} = w_t - \frac{\text{learning rate}}{\sqrt{v_t + 1}} \cdot \text{gradient}, \tag{13.5}$$

where v_t is the sum of past squared gradients and ε is a minimal valued constant to avoid division by zero.

13.3.1.5.3 AdaDelta

AdaDelta [17] was introduced to address two key limitations of AdaGrad: the rapid decay of learning rates during training and the need to manually select a learning rate. As an extension of AdaGrad, AdaDelta modifies the way gradients are accumulated. Rather than summing all past gradients, it uses a fixed-size window to accumulate only the most recent gradients within that window.

The pseudocode for the AdaDelta algorithm is provided below:

Algorithm: AdaDelta

Require: Decay rate ρ, Constant ε

Require: Initial parameter x_1

1 : Initialize accumulation variables $E\left[g^2\right]_0 = 0, E\left[\Delta x^2\right]_0 = 0$

2 : for t = 1 : T do%% Loop over # of updates

3: Compute Gradient: g_t

4 : Accumulate Gradient: $E\left[g^2\right]_t = \rho E\left[g^2\right]_{t-1} + (1-\rho)g_t^2$

5 : ComputeUpdate: $\Delta x_t = -\dfrac{RMS\left[\Delta x\right]_{t-1}}{RMS\left[g\right]_t} g_t$

6 : Accumulate updates $E\left[\Delta x^2\right]_t = \rho E\left[\Delta x^2\right]_{t-1} + (1-\rho)\Delta x_t^2$

7 : Apply update $x_{t+1} = x_t + \Delta x_t$

8 : End for

13.3.1.5.4 RMSProp

RMSProp [18] was introduced by Geoffrey Hinton during Lecture 6a of his Neural Networks for Machine Learning course, which provided an overview of mini-batch GD. In Lecture 6e, he elaborated on RMSProp, explaining that it works by dividing the gradient by a running average of its recent magnitudes. Hinton also discussed RProp (resilient backpropagation), which relies solely on the sign of the gradient and adapts the step size individually for each weight. In contrast, RMSProp—short for root mean square propagation—tackles the challenge of adapting the learning rate for each weight by maintaining a moving average of the squared gradients.

The RMSProp algorithm uses a moving average of the squared gradient of each of the weights.

$$\text{Mean Square }(w, t) = 0.9 \quad \text{Mean Square }(w, t-1) + 0.1\left(\partial E/\partial w\ (t)\right)^2$$

The changes proposed in RMSProp are to divide the gradient by the square root of the mean squared, which makes the learning much better.

13.3.1.5.5 Adam (Adaptive Moment Estimation)

Adam is another important optimization algorithm using first-order gradient with minimal memory. Here, adaptive learning rates are calculated individually using different parameters, based on estimates of the first- and second-order gradient moments. Hence, the name Adam stands for adaptive moment estimation. Adam and other adaptive gradient methods can train stably even in extremely sharp regions of parameter space, while momentum-based GD tends to be pushed away from areas where the sharpness surpasses its stability limit. As a result, Adam can sometimes converge to significantly sharper solutions than momentum GD [19]. Adam is the amalgamation of two popular optimization methods [20]: AdaGrad, which uses sparse gradients, and RMSProp, which was proposed by Geoffrey Hinton. Adam offers several key advantages: the magnitude of parameter updates remains unaffected by gradient rescaling, step sizes are constrained by the step size hyperparameter, it performs well even when the objective function is non-stationary, and it is effective with sparse [20] gradients.

The Adam algorithm [20] for stochastic optimization proposed by Kingma from the University of Amsterdam, OpenAI, and Jimmy Lei Ba from the University of Toronto is as follows:

Algorithm: Adam

Require: α : Stepsize

Require: $\beta_1, \beta_2 \in [0,1)$: Exponential decay rates for the moment estimates

Require: $f(\theta)$: Stochastic objective function with parameters θ

Require: θ_0 : Initial parameter vector

$m_0 \leftarrow 0$ (Initialize first moment vector)

$v_0 \leftarrow 0$ (Initialize second moment vector)

$t \leftarrow 0$ (Initialize time step)

while θ_t not converged do

$t \leftarrow t+1$

$g_t \leftarrow \nabla_\theta f_t(\theta_{t-1})$(Get gradients w.r.t stochastic objective at time step t)

$m_t \leftarrow \beta_1.m_{t-1} + (1-\beta_1)g_t$(Update biased first moment estimate)

$v_t \leftarrow \beta_2.v_{t-1} + (1-\beta_2)g_t^2$(Update biased second raw moment estimate)

$\widehat{m_t} \leftarrow m_t/(1-\beta_1^t)$(Compute bias-corrected first moment estimate)

$\widehat{v_t} \leftarrow v_t/(1-\beta_2^t)$(Compute bias-corrected second raw moment estimate)

$\theta_t \leftarrow \theta_{t-1} - \alpha.\widehat{m_t}/\left(\sqrt{\widehat{v_t}} + \varepsilon\right)$(Update parameters)

end while

return θ_t(Resulting parameters)

RMSProp is an optimization algorithm closely related to Adam [18]. The main distinction between RMSProp with momentum and Adam lies in how they compute parameter updates: RMSProp with momentum applies momentum to the rescaled gradients, while Adam directly estimates the updates using running averages of the gradient's first and second moments.

13.3.1.5.6 AdaMax

AdaMax is a variant of the Adam optimizer that is based on the infinity norm. In Adam, weight updates are computed by scaling gradients inversely with respect to a (scaled) L^2 norm of their current and past gradients. This update rule can be generalized using the L^p norm, but it becomes numerically unstable for large values of p. Interestingly, when p approaches infinity, the result is a remarkably simple and stable algorithm known as AdaMax.

The algorithm is as follows:

Algorithm: AdaMax [20]

Require: α : Stepsize

Require: $\beta_1, \beta_2 \in [0,1)$: Exponential decay rates

Require: $f(\theta)$: Stochastic objective function with parameters θ

Require: θ_0 : Initial parameter vector

$m_0 \leftarrow 0$ (Initialize first moment vector)

$u_0 \leftarrow 0$ (Initialize the eponentially weighted infinity norm)

$t \leftarrow 0$(Initialize time step)

while θ_t not converged do

$t \leftarrow t + 1$

$g_t \leftarrow \nabla_\theta f_t(\theta_{t-1})$(Get gradients w.r.t. stochastic objective at time step t)

$m_t \leftarrow m_t . \beta_{t-1} + (1 - \beta_1) g_t$(Update biased first moment estimate)

$u_t \leftarrow \max\left(\beta_2 . u_{t-1}, |g_{t-1}|\right)$(Update the eponentially weighted infinity norm)

$\theta_t \leftarrow \theta_{t-1} - \left(\alpha / (1 - \beta_1^t\right).\left(m_t / u_t\right)$(Update parameters)

end while

return θ_t (resulting parameters)

13.3.1.5.7 AMSGrad

AMSGrad was introduced in response to observations by Reddi et al. [21], who found that the Adam optimizer failed to converge on certain datasets and was sometimes outperformed by SGD with momentum. To address this limitation, AMSGrad was proposed as a modified version of Adam that maintains the maximum of the squared gradients over time, rather than relying on their exponential moving average.

AMSGrad [21] is a variant of Adam that incorporates an exponential moving average. This method is designed to ensure convergence while retaining the advantages of both Adam and RMSProp. AMSGrad typically employs a smaller learning rate than Adam but maintains the idea of gradually reducing the influence of past gradients on the learning rate. The algorithm below outlines the pseudocode for the AMSGrad method.

Algorithm: AMSGrad

Input: $x_1 \in F$, step size $\{\alpha_t\}_{t=1}^{T}, \{\beta_{1t}\}_{t=1}^{T}, \beta_2$

Set $m_0 = 0, v_0 = 0$ and $\hat{v}_0 = 0$

for t = 1 to T do

$g_t = \nabla f_t(x_t), m_t = \beta_{1t} m_{t-1} + (1 - \beta_{1t}) g_t$

$\hat{v}_t = \max(\hat{v}_{t-1}, v_t)$ and $\hat{V}_t = \text{diag}(\hat{v}_t), x_{t+1} = \Pi_{F,\sqrt{V_t}}\left(x_t - \alpha_t m_t / \sqrt{\hat{v}_t}\right)$

end for

13.3.2 Second-Order Optimization Algorithms

13.3.2.1 Newton's Method

Newton's method is one of the popular numerical techniques for approximating the solutions of any real-valued function. This is an iterative process that refines the initial guess for the root by using the derivative of the function. This method achieves faster convergence compared to other methods, like the bisection method. In general, the objective of any optimization algorithm is to find an x-value such that the objective function f(x) is either minimized or maximized. Here, Newton's method uses both the first- and second-order derivatives of the function to iteratively refine the solution.

Newton's method [22] is given by the equation below:

$$x_{n+1} = x_n - \frac{f'(x_n)}{f''(x_n)}, n \geq 0, \tag{13.6}$$

where x_{n+1} is the next approximation and x_n is the current approximation. $f'(x)$ is the first derivative and $f''(x)$ is the second derivative of function f(x).

Generalizing the above equation to n-dimensions, we may replace the first-order derivative by the gradient ∇f and the second order derivative by the Hessian H [23], respectively.

$$H_f = \begin{bmatrix} \frac{\partial^2 f}{\partial x_1^2} & \frac{\partial^2 f}{\partial x_1 \, \partial x_2} & \cdots & \frac{\partial^2 f}{\partial x_1 \, \partial x_n} \\ \frac{\partial^2 f}{\partial x_2 \, \partial x_1} & \frac{\partial^2 f}{\partial x_2^2} & \cdots & \frac{\partial^2 f}{\partial x_2 \, \partial x_n} \\ \vdots & \vdots & \ddots & \vdots \\ \frac{\partial^2 f}{\partial x_n \, \partial x_1} & \frac{\partial^2 f}{\partial x_n \, \partial x_2} & \cdots & \frac{\partial^2 f}{\partial x_n^2} \end{bmatrix}$$

For n dimensions, the new iterative equation is thus written as

$$\mathrm{x}_{k+1} = \mathrm{x}_k - H(\mathrm{x}_k)^{-1} \nabla f(\mathrm{x}_k). \tag{13.7}$$

Newton's method may be interpreted geometrically as, at each iteration, the function f(x) is approximated by a quadratic function around the current point, and a step is then taken toward the maximum or minimum of this quadratic approximation. Newton's method typically converges to a local maximum or minimum much faster than GD. However, depending on the characteristics of the function and the accuracy of the initial guess, it can sometimes diverge. A limitation of Newton's method is that it requires computing both the first- and second-order derivatives of function f(x). However, obtaining the second-order derivative can sometimes be inconvenient and time-consuming.

13.3.2.2 Quasi-Newton Methods

Quasi-Newton methods are a class of iterative numerical techniques used to find the maximum or minimum of a function f(x), like Newton's method. However, unlike Newton's method, which requires the exact Hessian matrix (second-order derivatives), quasi-Newton methods approximate the Hessian using only first-order gradient information. This approximation is updated iteratively, providing an efficient alternative to directly computing and inverting the full Hessian. This makes quasi-Newton methods particularly well-suited for large-scale problems, where handling the full Hessian would be computationally intensive. In Newton's method, for every iteration, the Hessian is computed and inverted. But the quasi-Newton methods avoid the computation of the Hessian by approximating it with a positive matrix B, which is definite, and it gets iteratively updated by using information from previous steps [23]. The updating of matrix B has been left vague because the specific update for B is based on the specific quasi-Newton method used. However, all the quasi-Newton methods must share the fact that the Hessian approximation B must satisfy the quasi-Newton condition

$$B_{k+1}\left[x_{k+1} - x_k\right] = \nabla f\left(x_{k+1}\right) - \nabla f\left(x_k\right). \tag{13.8}$$

Quasi-Newton methods are extensively utilized across various domains, including ML for optimizing deep learning models, in optimization for solving nonlinear problems, and in engineering for tasks such as design optimization and control. The Broyden–Fletcher–Goldfarb–Shanno (BFGS) method is a widely used quasi-Newton technique that updates the Hessian approximation in a way that preserves its positive definiteness—an essential property for ensuring a descent direction during optimization [24].

13.3.2.3 Conjugate Gradient Descent

The conjugate gradient descent (CGD) method is a well-known iterative algorithm designed for solving large-scale unconstrained optimization problems. It is especially appreciated for its efficiency and simplicity, as it avoids the need to compute second-order derivatives, setting it apart from techniques like Newton's method. CGD method is popular for its simple algorithmic structure and low memory requirement, making it suitable for optimizing large-scale problems [25]. The CGD method [26] is more efficient in solving linear equations or optimizing a quadratic equation than the GD. Let us consider a system of linear equations $A\vec{x} = b$, where A is a symmetric positive definite matrix and b is a column of values. Here, A is symmetric positive definite, meaning $(\vec{x})^T A(\vec{x}) > 0$.

Let us consider $f(\vec{x}) = \frac{1}{2}(\vec{x})^T A(\vec{x}) - b^T(\vec{x})$, then $\nabla f(\vec{x}) = A\vec{x} - b$. So, minimal values of $f(\vec{x})$ are in fact solutions of $A\vec{x} - b = 0$.

The CGD method is a type of line search algorithm that ensures each new step does not reverse the progress made by previous steps. Unlike basic gradient ascent, it can optimize a quadratic function in fewer iterations. Specifically, if variable x has N dimensions (i.e., N parameters), the method can find the optimal solution in at most N steps. This is achieved by choosing each search direction to be conjugate to all previous directions, ensuring efficient progress toward the optimum as shown in Figure 13.9. For example, in a four-dimensional space, the method should reach the optimal point in no more than four steps.

The steps followed are as follows:

1. We begin the ascent in one direction.
2. For one direction, settle down at the optimal point.
3. Identify a new direction d_j that is A-conjugate to any of the previous moving directions d_i.

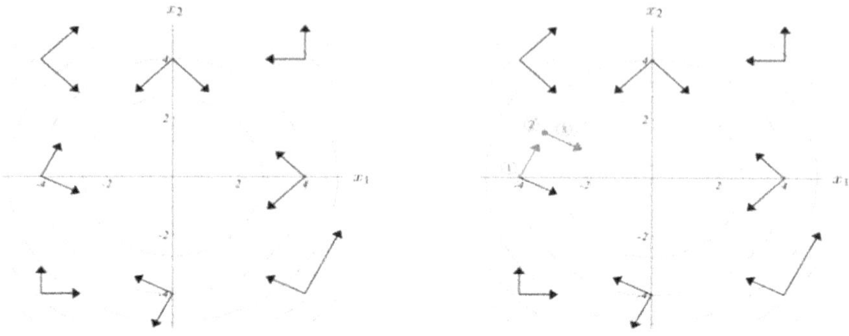

Figure 13.9 Conjugate gradient illustration.

(Source: [26])

Mathematically, as shown in the equation below, any new direction d_i must obey the A-conjugate with all previous directions $d_i : d_{(i)}^T A d_{(j)} = 0$, where A is the matrix in the quadratic equation.

The A-conjugate vectors should be independent of each other. Hence, for N-dimensions, N, A-conjugate vectors can span a space of N dimensions.

$$\mathrm{Ax}_* = \sum_{i=1}^{n} \alpha_i \mathrm{Ad_i}$$

Finding the values of $\propto$ and d is crucial for any of the conjugate gradient methods.

Algorithm:

Algorithm: Conjugate Gradient

$$r_0 = b - Ax_0; d_0 = r_0, \text{set k} = 0$$

$$\text{repeat } \alpha_k = \frac{r_k^T r_k}{d_k^T A d_k}, \; x_{k+1} = x_k + \alpha_k d_k, \; r_{k+1} = r_k - \alpha_k A d_k$$

If $\mathrm{r_{k+1}}$ is sufficiently small then exit loop

$$\beta_k = \frac{r_{k+1}^T r_{k+1}}{r_k^T r_k}, d_{k+1} = r_{k+1} + \beta_k d_k$$

$k = k + 1$

End repeat

The result is x_{k+1}

13.3.2.4 Trust Region Methods

Trust region methods (TRM) [27] begin by defining a neighborhood around the current best solution, within which a model can reasonably approximate the original objective function. Based on this model, TRM determines a step to move forward, guided by how well the model fits within this region. Unlike line search methods, which typically select a direction first and then determine the step size, TRM generally decides the step size before choosing the direction for improvement.

13.3.2.4.1 Trust Region Algorithm

First, we must find the trust region radius using the following:

Here, for every iteration, the parameter, trust-region $\Delta_k > 0$, is adjusted according to the agreement between the predicted model function m_k and the actual reduction in the function f as measured by the ratio

$$\rho_k = \frac{f(x_k) - f(x_k + p_k)}{f(x_k) - m_k(p_k)}. \tag{13.9}$$

The numerator denotes the actual reduction, and the denominator the predicted reduction. Always, the prediction will be non-negative. Hence, if ρ_k is negative, the step must be rejected. If there is good agreement between the model and approximation, i.e., $\rho_k \approx 1$, then Δ_k is increased. If the agreement is poor, i.e., ρ_k is small or negative, then Δ_k is decreased. The decision to accept step p_k is also based on ρ_k. Usually $x_{k+1} = x_k + p_k$, if $\rho_k \geq \sigma_0$, where σ_0 is small (typically, 10^{-4}); otherwise $x_{k+1} = x_k$. Based on the above ratio value, we change the trust radius for every iteration [27].

Here is the algorithm for TRM.

Given $\hat{\Delta} > 0$, $\Delta_0 \in (0, \hat{\Delta})$ and $\eta \in [0, 1/4)$:

For k = 0, 1, 2….

Obtain p_k (By solving the subproblem)

Evaluate ρ_k.

If $\rho_k < \frac{1}{4}$, $\Delta_{k+1} = \frac{1}{4}\Delta_k$.

Else,

if $\rho_k > \frac{3}{4}$ and $\|p_k\| = \Delta_k$, $\Delta_{k+1} = \min(2\Delta_k, \Delta)$; else, $\Delta_{k+1} = \Delta_k$.

If $p_k > \eta$, $x_{k+1} = x_k + p_k$; else, $x_{k+1} = x_k$.

13.3.2.4.2 Levenberg–Marquardt Algorithm

The Levenberg–Marquardt (LM) algorithm is another widely used optimization method that solves nonlinear least squares problems, blending the strengths of GD and Gauss–Newton techniques. It is especially effective in applications that involve model fitting to data, such as parameter estimation, ML, and inverse problems. The algorithm is renowned for its strong convergence capabilities and flexibility in handling different problem structures, making it a highly versatile tool in both theoretical research and practical applications. The algorithm calculates an approximate step using a modified Jacobian matrix along with a regularization parameter, which is dynamically adjusted based on the reduction ratio of the objective function.

Levenberg–Marquardt Algorithm [28]

Initialization

Choose the constants $\eta \in]0,1[, \mu_{\min} > 0$ and $\lambda > 1$. Select x_0 and $\mu_0 \geq \mu_{\min}$. Set $\gamma_0 = \mu_0 \|F(x_0)\|^2$ *and* $\bar{\mu} = \mu_0$.

For the values of $j = 0, 1, 2....$

1. Solve step (2), and let s_j denote such a solution.
2. Calculate $\rho_J = \dfrac{f(x_j) - f(x_j + s_j)}{m_j(0) - m_j(s_j)}$.
3. If $\rho_j \geq \eta$ then set $x_{j+1} = x_j + s_j$ and $\mu_{j+1} \in [\max\{\mu_{\min}, \bar{\mu}/\lambda\}, \bar{\mu}]$ and $\bar{\mu} = \mu_j$.
   Else, set $x_{j+1} = x_j$ and $\mu_{j+1} = \lambda \mu_j$.
4. Calculate $\gamma_{j+1} = \mu_{j+1} \|F(x_{j+1})\|^2$.

The LM algorithm is utilized across various fields, including image processing, where it enables precise and efficient parametric optimization in image mosaicking, leading to high accuracy and fast processing [29]. It is also applied in the sparse identification of dynamical systems, supporting parallel computing and streamlined model development, which makes it well-suited for real-time applications [30].

13.4 OPTIMIZATION ALGORITHMS FOR NEURAL NETWORKS

Neural networks are artificial models for computation developed to mimic biological neurons in the human brain. Their computational powers are very high in realizing many functions, primarily the logic functions. An artificial neural network (ANN) comprises several neurons arranged in vertical layers. The number of layers and the number of neurons are components of the architecture of an ANN. The first layer, where the input signals are received, is called the input layer, and there is at least one more layer called the output layer, where the neurons generate the output signals. The set of all neurons from one layer is connected to all the neurons in the next layer through directed lines and is attached with weights. These weights contain information regarding the input signals. An ANN having only one input and one output layer of neurons is called a single-layer ANN. However, besides these two layers, there may be one or more layers in between, called the hidden layers. The power of an ANN depends upon the number of neurons and their interconnections.

13.4.1 Deep Neural Networks

A DNN is a variant of an ANN with several hidden layers [15]. These networks are called "deep" because they have several hidden layers that process data, extracting increasingly complex features at each layer. DNNs can model complex, nonlinear relationships. They are capable of feature learning, automatically identifying patterns in data without manual intervention. DNNs are commonly used in tasks like image recognition, natural language processing, and autonomous systems [31].

13.4.2 Layer-Wise Adaptive Rate Scaling

Layer-wise adaptive rate scaling (LARS) [32] is an optimization method specifically designed for training DNNs like convolutional networks, particularly on large-scale datasets and models. It addresses the challenges of training very deep networks or networks with layers that vary greatly in their parameter scales.

Scaling Learning Rates Per Layer: Instead of applying a single global learning rate to all parameters, LARS adjusts the learning rate for each layer individually. This is done by computing a scaling factor for each layer based on the magnitude of its weights and gradients. The goal is to ensure that the updates to the parameters are proportional across layers, improving convergence speed.

LARS is particularly effective in training large models on distributed systems (e.g., using graphical processing units or tensor processing units). It helps avoid instability caused by differences in layer scales during training. Also, it promotes faster and more stable convergence of the optimization process compared to traditional methods like SGD. LARS is often used in combination with other optimization methods like SGD to improve training performance. It is widely applied in large-scale models like ResNet or in fields like computer vision.

For each layer, the learning rate is adjusted by introducing a layer-wise scaling factor.

The standard scaling factor is done by using $w \leftarrow w - \eta \nabla L$, where w is the weight of the layer, ∇L is the gradient of the loss L with respect to w, and η is the learning rate.

Networks usually attach parameters to different layers with varying magnitudes, leading to updates being imbalanced and a varied speed of updating in different layers. However, as the learning rate is scaled in LARS based on the magnitude of the weights and gradients, this problem does not arise.

The scaling factor λ_i is computed by using the following formula:

$$\lambda_i = \frac{\|w_i\|}{\|g_i\| + \varepsilon}, \tag{13.10}$$

where w_i is the weight of the ith layer, g_i is the gradient of the loss with respect to w_i, and ε is a small constant to avoid the possibility of division by zero.

The updating rule is adjusted for the ith layer as $w_i \leftarrow w_i - \eta\lambda_i g_i$.

This factor λ_i scales the global learning rate η for each layer individually, ensuring consistent updates. In combination with standard optimizers, line of the SGD, LARS is applied.

13.4.3 Lookahead Optimizer

The lookahead optimizer [33] is a novel optimization algorithm that iteratively updates two sets of weights, enhancing the learning stability and reducing the variance of its inner optimizer, such as SGD or Adam. The purpose of this algorithm is to find the best possible parameters for a neural network or ML model to minimize the loss function. The optimizer determines how the model's weights should be updated during training so the model can learn to make accurate predictions. It enhances optimization by incorporating a two-step process: managing fast weights and slow weights. Fast weights are updated frequently using any standard optimizer like SGD or Adam. Slow weights "look ahead" to the trajectory of fast weights and adjust themselves periodically based on the position of the fast weights.

The fast weights represented by θ_t are updated frequently using any base optimizer (e.g., SGD and Adam), and the updating formula is given by

$$\theta_{t+1} = \theta_t - \eta \nabla L(\theta_t), \tag{13.11}$$

where, $\theta_t \rightarrow$ Fast weights at iteration t, $\eta \rightarrow$ Learning rate, and $\nabla L(\theta_t) \rightarrow$ Gradient of the loss function L with respect to the fast weights.

After k fast weight updates, the slow weights ϕ_t are updated using the interpolation

$$\phi_{t+1} = \phi_t + \alpha\left(\theta_t - \phi_t\right), \tag{13.12}$$

where $\phi_t \rightarrow$ slow weights at iteration t, α is the hyperparameter controlling interpolation (usually $0 < \alpha < 1$), and θ_t are the current fast weights. This step helps ensure smoother transitions and avoids overshooting or oscillations during optimization.

Lookahead ensures that the optimization trajectory remains stable by periodically aligning the slow weights with the fast weights. Over iterations, the model converges to a point where

$$\phi^* = \theta^* .$$

The slow weights smooth out rapid fluctuations from fast weight updates, improving convergence stability. The updates are stabilized by regularly interpolating, and the optimizer avoids local minima and oscillations, leading to better generalization across unseen data. It works with base optimizers like SGD or Adam, inheriting their convergence guarantees.

13.4.4 Applications of Gradient-Based Methods

In the current high-dimensional data environment and advanced ML techniques, the gradient-based methods have applications in various popular domains. Some of the widely implemented domains are as follows. Aerodynamic Optimization: In this application, gradient-based methods are used to develop efficient aerodynamic designs, balancing computational cost with high fidelity analysis [34]. Interference Suppression: In the case of linear arrays, the gradient-based optimization maximizes main beam gain and minimizes sidelobe levels. Gradient-based optimization algorithms are powerful, with limitations. The major limitations are faced while handling nonlinearity and multimodality in objective functions, which necessitate the use of heuristic and metaheuristic algorithms in complex scenarios [35]. Also, the other major concern is with the computational costs of gradient evaluation in high-dimensional problems, which prompts the budding researchers in the development of advanced gradient methods to mitigate these challenges [34].

13.5 CONCLUSION

To conclude, in this chapter, we introduced the basics of ML and various types of ML algorithms. Also, we shifted our focus to the optimization algorithms, linked to the optimization of the error function or loss function in ML tasks, which is very much essential for any ML engineer. Also, we explored various gradient-based optimization methods, particularly first-order and second-order optimization algorithms, along with their application related to ML model use cases. With this background on the math behind the optimization algorithms and the context, ML practitioners or researchers can generate informed decisions that impact the training process significantly in ML applications. Finally, in the current context, with the increased complexity of ML models, mastering these optimization algorithms aids ML practitioners and researchers in theoretical research and real-world use-case implementation.

REFERENCES

1. Karthick, K. 2024. Comprehensive overview of optimization techniques in machine learning training. *Control Systems and Optimization Letters* 2(1): 23–27. https://doi.org/10.59247/csol.v2i1.69
2. Sun, S., Cao, Z., Zhu, H., and Zhao, J. 2020. A survey of optimization methods from a machine learning perspective. *IEEE Transactions on Systems, Man, and Cybernetics* 50(8): 3668–3681. https://doi.org/10.1109/TCYB.2019.2950779
3. Gzar, D., Mahmood, A.M., and Abbas, M.K. 2022. A comparative study of regression machine learning algorithms: Tradeoff between accuracy and computational complexity. *Mathematical Modelling of Engineering Problems* 9(5): 1217–1224. https://doi.org/10.18280/mmep.090508
4. Filali Razzouki, O., Charroud, A., El Allali, Z., Chetouani, A. and Aslimani, N. Dec 2024. A survey of advanced gradient methods in machine learning, pp. 1–7. https://doi.org/10.1109/commnet63022.2024.10793249
5. Maurya, M.R., and Yadav, N. 2023. *A Comparative Analysis of Gradient-Based Optimization Methods for Machine Learning Problems*. Springer International Publishing, Berlin, pp. 85–102. https://doi.org/10.1007/978-981-99-3432-4_7
6. Sakovich, N., Aksenov, D., Плешакова, E. B., and Gataullin, S. 2024 Sep. MAMGD: Gradient-based optimization method using exponential decay. *Technologies (Basel)* 12(9): 154. https://doi.org/10.3390/technologies12090154
7. Tian, X. 2024 Jun. Adaptive energy-based gradient methods for large-scale optimization and data-driven discovery of dynamical systems via neural networks. https://doi.org/10.31274/td-20240617-16
8. Gao, R., Yi, T., Qi, W., and Liu, S. 2021 Sep. Gradient-based optimization method for interference suppression of linear arrays by the amplitude-only and phase-only control. *International Journal of Microwave and Wireless Tech*nologies, 1–7. https://doi.org/10.1017/S1759078721001355
9. Gu, Y., and Chen, X. 2024 Oct. Towards differentiable multilevel optimization: A gradient-based approach. https://doi.org/10.48550/arxiv.2410.11312
10. Andel Finger, P. 2021 Mar. Differentiable agent-based simulation for gradient-guided simulation-based optimization. *Learning* [Online]. Available: https://arxiv.org/abs/2103.12476
11. Tripathy, B.K., and Anuradha, J. 2015. *Soft Computing—Advances and Applications*. Cengage Learning publishers, New Delhi, ISBN-9788131526194 (*)
12. Ruder, S. 2016. An overview of gradient descent optimization algorithms. arXiv preprint arXiv:1609.04747
13. Qian, N. 1999. On the momentum term in gradient descent learning algorithms. *Neural Networks* 12(1): 145–151. https://doi.org/10.1016/S0893-6080(98)00116-6
14. Razzouki, O.F., Charroud, A., El Allali, Z., Chetouani, A., and Aslimani, N. 2024. A survey of advanced gradient methods in machine learning, pp. 1–7. https://doi.org/10.1109/commnet63022.2024.10793249
15. Bhattacharyya, S., Snasel, V., Hassanian, A.E., Saha, S., and Tripathy, B.K. 2020. *Deep Learning Research with Engineering Applications*. De Gruyter Publications, Berlin, ISBN: 3110670909
16. Duchi, J., Hazan, E., and Singer, Y. 2011. Adaptive subgradient methods for online learning and stochastic optimization. *Journal of Machine Learning*

Research 12(61): 2121–2159. http://digitalassets.lib.berkeley.edu/techreports/ucb/text/EECS-2010-24.pdf

17. Zeiler, M.D. 2012. *Adadelta: An adaptive learning rate method*. arXiv preprint arXiv:1212.5701.
18. Hinton, G., Srivastava, N., and Swersky, K. 2012. Neural networks for machine learning lecture 6a overview of mini-batch gradient descent. *Cited on* 14(8): 2.
19. Cohen, J.M., Ghorbani, B., Krishnan, S., Agarwal, N., Medapati, S., Badura, M., and Gilmer, J. 2022. *Adaptive gradient methods at the edge of stability*. arXiv preprint arXiv:2207.14484.
20. Kingma, D.P., and Ba, J. 2014. *Adam: A method for stochastic optimization*. arXiv preprint arXiv:1412.6980.
21. Reddi, S.J., Kale, S., and Kumar, S. 2019. *On the convergence of Adam and beyond*. arXiv preprint arXiv:1904.09237.
22. Bhuyan, M.H. 2013. Newton's numerical method for optimization: A simpler approach. *Daffodil International University Journal of Science and Technology* 8(1): 17.
23. Lam, A. 2025. BFGS in a nutshell: An introduction to quasi-Newton methods. *Towards Data Science*. https://towardsdatascience.com/bfgs-in-a-nutshell-an-introduction-to-quasi-newton-methods-21b0e13ee504
24. Kumar, S., Mahato, N.K., Ansary, M.A.T., and Ghosh, D. 2023. A quasi-Newton method for uncertain multiobjective optimization problems via robust optimization approach. https://doi.org/10.48550/arxiv.2310.07226
25. Alhawarat, A., Alhamzi, G., Masmali, I., and Salleh, Z. 2021. A descent four-term conjugate gradient method with global convergence properties for large-scale unconstrained optimisation problems. *Mathematical Problems in Engineering*, 1–14. https://doi.org/10.1155/2021/6219062
26. Hui, J. 2018. RL—Conjugate gradient. *Medium*. https://jonathan-hui.medium.com/rl-conjugate-gradient-5a644459137a
27. Khare, S. 2021. A hands-on blog on trust region methods (with mathematical intricacies!). *Medium*. https://medium.com/trust-region-methods/temp-blog-1b51189594a
28. Bergou, E.H., Diouane, Y., and Kungurtsev, V. 2020. Convergence and complexity analysis of a Levenberg-Marquardt algorithm for inverse problems. *Optimization and Control*. arXiv: https://arxiv.org/abs/2004.03005
29. Yanjun, F., Kuntao, Y. and Wendong, Z., 2007. Image mosaic based on Levenberg-Marquardt algorithm. *Laser Journal* 28(5): 46–48.
30. Haring, M., Grotli, E.I., Riemer-Sorensen, S., Seel, K. and Hanseen, K.G., 2022. A Levenberg-Marquardt algorithm for sparse identification of dynamical systems. 2022. *IEEE Transactions on Neural Networks and Learning Systems* 34(11): 9323–9336. https://doi.org/10.1109/tnnls.2022.3157963
31. Tripathy, B.K., Parikh, S., Ajay, P., and Magapu, C. 2022. Brain MRI segmentation techniques based on CNN and its variants (Chapter-10). *Brain Tumor MRI Image Segmentation Using Deep Learning Techniques* (Editor: J. Chaki). Elsevier Publications, Amsterdam, pp. 161–182. https://doi.org/10.1016/B978-0-323-91171-9.00001-6
32. Ginsburg, B., Gitman, I., and You, Y. 2018. *Large Batch Training of Convolutional Networks with Layer-wise Adaptive Rate Scaling*. https://openreview.net/pdf?id=rJ4uaX2aW

33. Zhang, M.R., Lucas, J., Hinton, G.E., and Ba, J. 2019. Lookahead Optimizer: k steps forward, 1 step back. *Learning*. arXiv: https://arxiv.org/abs/1907.08610v1
34. Liu, M., and Qian, R. 2024 May An efficient aerodynamic optimization method based on approximate gradient analysis. https://doi.org/10.1088/1742-6596/2764/1/012081
35. Asadi, S., Gharibzadeh, S., Zangeneh, S., Reihanifar, M., Rahimi, M., and Abdullah, L. 2024, Sep. Comparative analysis of gradient-based optimization techniques using multidimensional surface 3D visualizations and initial point sensitivity. https://doi.org/10.48550/arxiv.2409.04470

Index

For Product Safety Concerns and Information please contact our EU representative GPSR@taylorandfrancis.com
Taylor & Francis Verlag GmbH, Kaufingerstraße 24, 80331 München, Germany

www.ingramcontent.com/pod-product-compliance
Ingram Content Group UK Ltd.
Pitfield, Milton Keynes, MK11 3LW, UK
UKHW020434270926
475702UK00051B/494
* 9 7 8 1 0 4 1 0 7 7 6 1 9 *